"Zavala and Aguirre should be commended for their take on what it means to be a culturally responsive mathematics teacher. This book positions mathematics teachers as identity workers. The vignettes and teaching stories bring the Culturally Responsive Mathematics Teaching Tool to life. Zavala and Aguirre graciously share their thoughts about how to implement this framework in a variety of settings, including elementary, special education, and multilingual contexts."

–Pam Seda and Kyndall Brown
Educational Consultants
Los Angeles, CA

"Joyful, humanizing, rich mathematics teaching and learning come alive in this book. Zavala and Aguirre share a compelling framework for culturally responsive mathematics teaching that will fuel your imagination and motivate you to grow your teaching in new ways. With this book, you and your colleagues will have the power to transform your students' mathematics experiences."

–Elham Kazemi
Professor, Mathematics Education,
University of Washington
Seattle, WA

"Written in flowing narrative and interspersed with the voices of students and teachers, this book is a must-read for math educators committed to equity using culturally responsive mathematics instruction. The planning tool for Culturally Responsive Mathematics Teaching supports every step of instructional design of math to reach and teach each and every child in our mathematics classrooms."

–Tonya Bartell
Associate Professor of Mathematics Education
Associate Director of Elementary Programs,
Michigan State University
East Lansing, MI

"*Cultivating Mathematical Hearts* is a love letter to children, families, communities, and teachers. Both practical and inspirational, this book offers the perfect blueprint with the rationale, practices, lesson design tools, and classroom examples to ensure culturally responsive mathematics teaching is available for all. Thank you to the authors for sharing your heart work—your vision, expertise, and radical hope!"

–Cathery Yeh
Assistant Professor of Curriculum and Instruction
Asian American Studies Core Faculty, The University of Texas at Austin
Austin, TX

"The authors of *Cultivating Mathematical Hearts* skillfully escort us on a journey in which they define, unpack, and offer profound examples of Culturally Responsive Mathematics Teaching. This book is complex, practical, and personal. New and experienced educators are guaranteed to recommit to centering love, joy, curiosity, humanity, and community in mathematics teaching. The authors end with a powerful yet empathetic call to action. This book is essential on our collective journeys to liberation."

–Marrielle Myers
Associate Professor of Mathematics Education,
Kennesaw State University
Atlanta, GA

"This book is inspiring and essential reading for everyone who advocates for children and wants them to thrive! The Culturally Responsive Mathematics Teaching Tool will guide teachers, teacher educators, and administrators who want to pursue meaningful and humane mathematics teaching. The stories, instructional examples, and activities will promote thoughtful reflections and discussions as well as unify efforts in schools, districts, and communities to facilitate engagement, joy, hope, and justice."

–Kathryn Chval
Dean and Professor, Mathematics Education,
University of Illinois Chicago College of Education
Chicago, IL

"No matter what stage we are in in our understanding of Culturally Responsive Mathematics Teaching, the authors brilliantly guide us on how to transform our classrooms by acknowledging, taking action, and holding ourselves accountable. We have the power to retool pedagogy that centers the students', families', and communities' conocimientos (knowledge) and ways of doing mathematics. Exemplary cases from the field are included to illustrate how this work is done del corazón (from the heart)."

–Sylvia Celedón-Pattichis
Professor, University of Texas at Austin
Austin, TX

CULTIVATING MATHEMATICAL HEARTS

Culturally Responsive Mathematics Teaching in Elementary Classrooms

MARIA DEL ROSARIO ZAVALA
JULIA MARIA AGUIRRE

A SAGE Publishing Company

For information:

Corwin
A SAGE Company
2455 Teller Road
Thousand Oaks, California 91320
(800) 233–9936
www.corwin.com

SAGE Publications Ltd.
1 Oliver's Yard
55 City Road
London, EC1Y 1SP
United Kingdom

SAGE Publications India Pvt. Ltd.
Unit No 323–333, Third Floor, F-Block
International Trade Tower
Nehru Place
New Delhi – 110 019
India

SAGE Publications Asia-
Pacific Pte. Ltd.
18 Cross Street #10–10/11/12
China Square Central
Singapore 048423

Vice President and Editorial Director:
    Monica Eckman
Associate Director and Publisher,
    STEM: Erin Null
Senior Editorial Assistant:
    Nyle De Leon
Production Editor: Tori Mirsadjadi
Copy Editor: Michelle Ponce
Typesetter: Integra
Proofreader: Lawrence W. Baker
Indexer: Integra
Cover Designer: Scott Van Atta
Marketing Manager:
    Margaret O'Connor

Printed in Canada.

Paperback ISBN 9781071850107

This book is printed on acid-free paper.

MIX
Paper from
responsible sources
FSC® C103567

23 24 25 26 27   10 9 8 7 6 5 4 3 2 1

CONTENTS

Visit corwin.com/books/mathematical-heart-277718
for a full-color version of the CRMT2 and the free
Book Study Guide.

PREFACE

We want to start by saying thank you! If you are reading this book, then you have already made a commitment to enhance your math teaching knowledge and practice that center children, families, and communities in meaningful mathematics learning and teaching. This commitment is part of the heartwork needed to cultivate joy and justice in the elementary mathematics classroom.

Ten years ago, we started this journey together to create conceptual and practical tools to help preservice and inservice teachers to be more culturally responsive in mathematics teaching. At that time, the emphasis on student thinking and mathematical discourse did not fully affirm the multitude of mathematical strengths, languages, creativity, critical thinking, and resources students brought from their home and communities. In our own scholarship and teaching, we needed different types of tools that would be comprehensive enough to respect the complexity of teaching and accessible enough to help teachers engage in purposeful pedagogical design work and reflection in their practice with an equity lens. In 2013, we published a journal article entitled "Making Culturally Responsive Mathematics Teaching Explicit: A Lesson Analysis Tool" (Aguirre & Zavala, 2013). In that article, we defined Culturally Responsive Mathematics Teaching (CRMT) and created a multidimensional lesson analysis tool that encompassed five dimensions associated with math reform (intellectual support, depth of knowledge, math analysis, math discourse, and student engagement) and four dimensions related to equity (academic language support for English language learners, use of English as a second language scaffolding strategies, funds of knowledge/cultural and community support, and use of critical knowledge and social justice). We had a tool that took the best of reform math practices and paired it with practices that emphasized attention to language, culture, and critical math knowledge. We felt this tool captured complexity in productive ways. In addition, we've been contacted by many scholars and teachers who have used this tool in professional development settings, math methods courses, and reflecting on their own instruction.

However, as our math education work continued within the sociopolitical turn of mathematics education (Gutiérrez, 2010/2013), and we collaborated with more teachers working with culturally and linguistically diverse children in elementary settings, we realized that our tools also needed to evolve to cultivate and sustain meaningful mathematical learning spaces for children—spaces that affirmed their whole beings and helped them learn.

Cultivating Mathematical Hearts: Culturally Responsive Mathematics Teaching in Elementary Classrooms is a love letter to all the children, families, communities, and teachers we've worked with to make mathematics more meaningful and humanizing. It is written to re-center mathematics as a humanizing endeavor, one that has always been a part of us, like our heart; something that we can strengthen and is needed to help us better understand and make positive change in our world.

Our approach to this book is deeply connected to who we are as mothers, scholars, teachers, and activists. Our positionality matters as two Latina math teacher educators with unapologetic commitments to social justice in mathematics education. We are friends, sisters of the heart, and collaborators on this journey to normalize culturally responsive mathematics teaching in the classroom.

However, acting on our commitments can be challenging, especially while navigating the twin pandemics of COVID and systemic racism. COVID disproportionately impacted communities of color and working-class communities leaving many coping with trauma while building up resiliency. Teachers, like other essential workers, dug deep to support children and their families, while also trying to keep themselves and their families safe. COVID also set off a panic during which children, especially from under-resourced communities, suffered tremendous learning loss that needed remediation rather than innovation. At the same time, systemic racism continues to affect how we live our lives. The murders of Breonna Taylor and George Floyd at the hands of police in 2020 catalyzed the Black Lives Matter movement to continue to fight for justice and to end state-sanctioned brutality. Efforts to end anti-Black, anti-Asian, and anti-Immigrant attacks that fuel violence in our communities must continue. We must demand justice for the missing and murdered Indigenous women of our communities. These fights for justice are being replicated across the globe as systemic racism is not just a U.S. phenomenon but is also endemic around the world. Elementary classrooms are not immune from these social forces. Children are watching, listening, and taking action. We stand with antiracist and social justice statements that call for change:

> *An antiracist position in mathematics education is a pledge to dismantle systems and structures that maintain racism within teaching and learning mathematics from challenging belief systems that perpetuate microaggressions to disrupting the role mathematics classes play in pushing students out of schooling. (p. 2, TODOS, 2020)*

We invite you to stand with us because children and their humanity are at the heart of culturally responsive mathematics teaching.

WHO IS THIS BOOK FOR?

This book is for everyone who believes there must be a better, stronger way to teach mathematics that fosters student joy and curiosity, rather than fear and disconnection; a way that demonstrates to children that mathematics is both a mirror and a lens—it reflects who you are and who you are becoming, and it helps you analyze the world around you. It offers a way to see mathematics as a tool to make positive change in our communities and build foundational knowledge and skills to solve novel and complex problems in our world and its future.

Wherever you are in your teaching career, this book is for you. This book is for newer teachers who are both excited and overwhelmed at the prospect of teaching multiple subjects to young children or who have had a rocky relationship with mathematics and do not want to pass on negative views of mathematics to their future students. This book is also for more experienced teachers, instructional coaches, and teacher leaders who want to meet the needs of students who may or may not like mathematics or who definitely don't see it connecting to their lives. In addition, whether you love or hate your mathematics curriculum, you will find useful ideas here. This book is for all of us grappling (with)in a mathematics education system that is inequitable and restrictive at this time. We want this book to facilitate creative freedom in mathematics teaching and learning and unite our elementary mathematics education community to make the change we want to see.

We also understand that change is hard. It requires courage, patience, and practice. We're offering something different—not an add-on to your current teaching but rather a way to refresh your approach to mathematics instruction and maybe even heal your own mathematical trauma. Our book provides a wide range of teaching examples across different elementary classroom settings that support culturally, linguistically, and neurodiverse mathematical learners. Along the way, we have activities designed to deepen and develop your own mathematical identity, as well as redefine and refine your classroom instruction. We will also show how you can retool your practice, through cycles of design, analysis, and reflection. We hope that the ideas, activities, tools, and stories presented in this book will free up space for you to be more creative, innovative, and reflective with your mathematics teaching and for building stronger connections with your amazing students and their families and communities.

ACKNOWLEDGING THE SOCIOPOLITICAL CONTEXT

The sociopolitical context of mathematics education means we recognize that mathematics is always political and socially constructed (Gutierrez, 2010/2013). We identify with particular social categories (e.g., Latina, Chicana, multilingual, multiethnic, biracial) that have been historically marginalized in societal and school settings. We recognize that these social categories have evolving meanings and histories. They are negotiated. They are personal and political. In this book, various terms are used to describe people. For example, we used categories like "Black" and "African American" to express diverse ways group members self-identify and are identified by others as connected to the African diaspora. We also use "Latinx" as a pan-ethnic label that encompasses people with ancestry from cultures located south of the United States, including Mexico, countries in central and south America, and the Caribbean. It also challenges the gender binary associated with Latina/o. We use "Native American," "First Nations," and "Indigenous" to acknowledge the first peoples of the land and tribal sovereignty. We also recognize that categories such as "Asian" and "white" include multiple identifications. So, wherever possible, we try to identify specific cultural and ethnic identities. In addition, we want to affirm multilingualism by centering on learning more than one language using terms such as "multilingual learner" or "bilingual learner." Using the term "English language learner" often elevates the acquisition of English at the expense of a person's home languages. In this book, we offer illustrations of students engaged in "translanguaging," which means fluidly using their linguistic resources to make meaning and communicate with each other (Garcia, 2009). We also use the term "neurodivergent" to move away from "learning disability" labels often associated with students receiving special education services. Some exceptions to our language choices will be evident when we cite relevant literature or situation terms within broader policy or research discussions, as may be expected when bringing different viewpoints together. However, we have done our best to be consistent. We invite readers to critically reflect on the social categories you use in your own life and work in schools, especially discourses related to mathematics.

We also make some words plural that readers may not be familiar with seeing in this way. For example, we pluralize "knowledge" to "knowledges" to show how sometimes a singular sense of a concept means there is a legitimate, single form, implying all else is illegitimate. We hope to disrupt these power binaries by acknowledging that all communities and families have their forms of knowledge (i.e., their "knowledges") and contain multitudes of pluralistic identities (i.e., we are more than one singular identity). Therefore, when you see words like knowledges, logics, identities, and other terms that may usually be singular you should know we intentionally selected the plural form.

THE BOOK'S ORGANIZATION

This book is organized into two main parts. Part 1 describes the fundamental principles, ideas, and vision (i.e., foundations) of CRMT. It also introduces a multidimensional framework for CRMT. Chapter 1 provides the purposes and principles that inform CRMT. Understanding the origins of this work provides a solid foundation for your journey. Chapter 2 introduces the

Culturally Responsive Mathematics Teaching Tool (CRMT2) including how it is structured into three strands: Knowledges and Identities; Rigor and Support; and Power and Participation. The rubric style tool will be described along with ways it can be used in practice. Think of it as a map to help guide you on your journey. Chapters 3 through 5 dive deep into each strand and the accompanying dimensions. Teaching stories, instructional examples, and try-it activities give readers explicit examples of CRMT in action and how they can get started with this work. Each strand chapter also provides opportunities to critically reflect. Interspersed throughout each chapter are "pause and reflect" questions. A set of discussion questions conclude each chapter. Chapter 3 focuses on the strand Knowledges and Identities. This strand has three dimensions: *Centering Cultural and Community Funds of Knowledge*; *(Re) Humanizing Mathematics*; and *Honoring Student Thinking and Ideas*. The Knowledges and Identities strand emphasizes the unequivocal faith we have in students as complex and creative human beings. Chapter 4 focuses on the strand Rigor and Support. The three dimensions of this strand are *Sustaining High Cognitive Demand*; *Scaffolding Up*; and *Affirming Multilingualism*. This strand supports student empowerment as thinkers, doers, and communicators of mathematics. Chapter 5 focuses on the strand Power and Participation. This strand also has three dimensions: *Distributing Intellectual Authority*; *Disrupting Status and Power*; and *Analyzing and Taking Action*. These dimensions reflect a justice orientation about who can do mathematics, whose voices matter in mathematics learning spaces, and how mathematics can be used to challenge inequities and injustices in the world.

Part 2 features a set of chapters that illustrate CRMT in action. These chapters (6–9) emphasize teacher voice as they are written by elementary teachers and instructional coaches detailing how they've used the CRMT tool (CRMT2) to plan, enact, and reflect on their mathematics teaching practice. The chapters are grounded in different elementary classroom settings using examples of student work and participation to illustrate CRMT. Chapter 6 focuses on how the CRMT2 helped a first-grade teacher and instructional coach adapt a district-mandated math lesson to be more culturally responsive. Chapter 7 focuses on how the CRMT2 provided a critical lens to help a teaching partnership (fourth-grade teacher and instructional coach) describe the evolution of their math teaching practice to be more culturally responsive over four years. Chapter 8 describes how CRMT is enacted in a bilingual classroom that emphasizes mathematical thinking, doing, and communicating with multiple modalities, including translanguaging. Chapter 9 describes CRMT in special education settings that helps to showcase mathematical strengths evident, but often overlooked, in this learning space.

Our hope for this book is to provide educators a comprehensive resource that will enhance mathematics teaching and learning in the elementary classroom, especially for children historically marginalized and segregated in schools (e.g., children of color, multilingual learners, and children receiving special education services). We invite you to use this book in any way that will help you reflect and grow your math teaching practice. It might help to start with Part 1 to deepen your own understanding of CRMT. Then for specific illustrations of how teachers have utilized CRMT in elementary classroom settings, you may want to read a chapter from Part 2 that connects to your current role or interest. All guest author chapters in Part 2 provide detailed accounts of mathematics teaching and learning. We encourage you to read all the chapters but perhaps start with the one closest to your mathematical heart. If you are reading this as a book study or with colleagues, you can find a Book Study Guide here: **corwin.com/books/mathematical-heart-277718**.

Thank you again for your courage and commitment to learning more about CRMT. We hope this book will give you freedom to deepen your faith in the humanity of children and all they bring to the math learning space, support student empowerment with and through mathematics, and cultivate joy and justice with mathematics in elementary classrooms.

ACKNOWLEDGMENTS

We would like to first thank all the amazing children, families, and educators we've worked with over the past several years who continue to inspire us with their brilliance, agency, action, and love. This work is rooted in ancestral knowledge, ways of being and ways of learning that we want all to help reclaim as part of our humanity—our hearts. Thank you to those who stand with us on this justice journey. This path is never easy, yet it can be joyful and illuminating with new insights and ideas as we move forward together.

We would like to thank our wonderful guest authors and extraordinary educators whose criticality and sociopolitical consciousness of mathematics education leads to cultivating math learning spaces where children thrive. Thank you, Melissa Adams Corral, Olivia Canning, Kaitlin Kaplewicz, Talya Kemper, and Holly Tate for giving this book personal portraits of the possible with the students and families you work with. Thank you for taking the time to share your own journey with Culturally Responsive Mathematics Teaching (CRMT).

We would also like to thank scholars and educators who helped us hone the ideas for this work. Special appreciation to the EQ-STEMM project collaborators (Advancing Equity and Strengthening Teaching Through Elementary Mathematical Modeling, www.eqstemm.org): Mary Alice Carlson, Jennifer Suh, and Erin Turner. Your friendship and insight have helped make this work better. Thank you to the math educator-graduate students who work closely with us on CRMT work, especially Elzena McVicar and Holly Tate.

We would also like to thank the numerous named and unnamed colleagues in mathematics education and beyond who encouraged us, gave us feedback on the framework over time, and were excited we were writing this book. This book has been many years in the making, and your support and constant refrain of how important it was for us to get this done have been a lifeline.

Thank you to Corwin for making this book possible. Special thanks to Erin Null, who was open to this book idea and provided insightful feedback and guidance from start to finish. Thank you for your patience and persistence.

We would also like to thank Nyle DeLeon and Jessica Vidal, who provided logistical support through this process. We believe this book will make a tremendous contribution to the field. Corwin's supportive collaboration was essential and deeply appreciated.

We would also like to thank the artists who contributed artistry to this book. Míl gracias to the Northwest Chicano artist and activist Jake Prendez for the beautiful piece, *7 Generations of Genetic Memory*. You can find his work at Nepantla Cultural Arts Gallery in Seattle, WA and online at https://www.jakeprendez.com/. Many thanks to Norm Mattox, whose poetry and love for mathematics and humanity light our path on this journey. His work can be found at https://www.mathmattox.com/.

Maria's Special Acknowledgments

First, I would like to thank Julia for being my teacher, collaborator, and friend for over twenty-one years! I could not imagine writing this book without you. You inspire me, you listen to my wild ideas and reground me with compassion and understanding, and you bring creative and new possibilities to how I think about my work on this planet. I am forever, forever-ever grateful to be living this life beside you.

Thank you to my mother, Susana Arbe, for your constant, unequivocal faith in me as a person. And for feeding me.

Thank you to my two children, who are now young but who I hope one day will read this dedication and understand just how profoundly being their mother has shaped me, personally and professionally. Ale and Yuca, you are amazing, and I'm honored to be your mama. Thank you for letting me run ideas past you, for sharing your mathematical thinking with me, even when the questions were kind of out of the blue. And because of your generosity, my student teachers also learn from your counting and problem-solving videos. Thanks for letting me share your brilliance.

Adam, I love you. Thank you for being my partner. Sometimes you are my rock, and sometimes I am your rock. And I would not have it any other way.

To my sister, brothers, and other family who I hold close and dear, thanks for nodding along politely when I explained what I do and also for the encouragement to keep being excellent. You all knew I loved math from the beginning, and you always thought it was cool. You're the best. Dad, I miss you, and I know I am making you proud, from one doctor to another.

Lastly, to the students and faculty at San Francisco State University. Because of your acceptance, curiosity, and commitment to social justice in education, I have felt nurtured to pursue and refine the ideas contained herein, even when we had difficult conversations or discourse where, ultimately, we did not agree. I am very appreciative of our community at SF State.

Julia's Special Acknowledgments

I would like to start by thanking Maria for this opportunity to collaborate on this book. We've known each other for twenty-one years, and it's been my honor to grow and learn from and with you. You have been a tremendous positive influence on cultivating my mathematical heart. You have helped sustain my critical equity and justice heartwork. I am forever grateful for our friendship.

I would also like to thank my family. Thank you, Jade, for being such a loving and patient spouse. Your tender heart and hand have gotten me through difficult times. I appreciate you making me laugh and being there when I'm not. Thank you for being on this journey with me. And to my children, Alesandra and Joaquín, you continue to inspire me as you move into adulthood with openness and compassion. I love you always. To my parents, Ricardo and Marlinda Aguirre. Thank you. My mother's faith in us knows no bounds. My father's spiritual presence is always with me whispering *Sí se puede* when I most need it.

I would also like to thank all of the preservice and inservice teachers I've had the honor to work with over the years. You continue to inspire me and push me to interrogate and innovate my heartwork. I have great faith in our next generation of educators to make a positive and lasting difference in the hearts and minds of our nation's young people.

ABOUT THE AUTHORS

 **Maria del Rosario Zavala** (she/ella) is the second-born child of Peruvian immigrant parents and the first of her family born in the United States. She is an associate professor of elementary education at San Francisco State University, specializing in mathematics and bilingual education. Dr. Zavala's work in education is informed by her multiple years of teaching and volunteering in K–12 classrooms and by raising two children with her math-teaching partner. Her research focuses on the mathematics identity development of children and how teachers deliver on the promise of Culturally Responsive Mathematics Teaching. She is particularly interested in how people, especially from historically marginalized populations, (re)claim their mathematical power.

 **Julia Maria Aguirre** (she/ella) is a biracial Chicana activist math education scholar and teacher educator. Her parents were educators and activists integrating faith, community, and justice growing up. Dr. Aguirre is a professor of education at the University of Washington Tacoma. Her work focuses on critical equity studies in mathematics education, teacher education, and culturally responsive mathematics pedagogy. She has taught mathematics in formal and informal classroom settings. A primary goal of her work is preparing new generations of teachers to make mathematics education accessible, meaningful, and relevant to today's youth. Dr. Aguirre welcomes all to join her in cultivating a more humanizing and just mathematics education.

Guest Authors

Holly Tate (she/her/hers) began her career as an elementary school teacher, before moving into a role as an instructional mathematics coach. After receiving her mathematics specialist degree, she began to work on her PhD, specializing in mathematics education and research methodologies. Holly's research interests include equity-focused mathematics teaching and learning, with special attention to how power and systems impact students' mathematical opportunities and identities. Holly centers collaborative learning alongside teachers and the school community. She is especially devoted to exploring the connection between mathematics and humanity.

Olivia Canning (she/her), originally from Long Island, New York, started her education career in Northern Virginia first as a special education teacher and later as a general education teacher working in various grades in K–5. Olivia has interests in using current research to improve student learning in both literacy and mathematics instruction and in creating a supportive inclusive classroom environment for students to feel safe, challenged, and part of a community. Olivia specializes in multisensory, research-based reading instruction to help struggling readers find success within the general education classroom.

Kaitlin (Katie) Kaplewicz (she/her/hers) began her career as a fourth-grade teacher before moving into her current role as a mathematics instructional coach at an elementary school. Early on in her career as a classroom teacher, Katie discovered her passion for equitable mathematics teaching and learning and decided to further her education by completing a graduate program to earn her mathematics specialist endorsement. She continues to find joy in inspiring both educators and students to experience excitement and curiosity on their journey with mathematics.

Melissa Adams Corral (she/ella) is an assistant professor of mathematics education at the University of Texas, Río Grande Valley. Before receiving her PhD, she spent seven years as a bilingual elementary teacher in central Texas. Her research deploys theories and methods from community organizing in classroom-based practice and explores topics shaping the intersection of race, language, and mathematics education.

 **Talya Kemper** (she/her) is an associate professor of education at Cal State University East Bay (CSUEB) as well as a teacher in the Oakland Unified School District. She has numerous classroom experiences as a paraprofessional and teacher of students with special needs and holds credentials in moderate to severe disabilities and physical and other health impairments in addition to her master's degree in education and her doctoral degree from the University of Washington. She has worked in teacher education both at CSU Chico and now CSUEB. Her research focus is on the use of alternative and augmented communication, particularly among emergent bilingual students and on issues of diversity in the design and research of educational interventions for students with autism. Dr. Kemper remains committed to her work in teacher training through CSUs and continues to teach in the classroom as a special education teacher.

PART 1

·················

FOUNDATIONS AND FRAMEWORK OF CULTURALLY RESPONSIVE MATHEMATICS TEACHING

Part 1 of this book provides fundamental ideas and examples of Culturally Responsive Mathematics Teaching (CRMT). It is important to understand what theories and scholarship ground this concept. This provides the foundation to launch your journey. We provide our working definition of CRMT, the conceptual grounding of this work, as well as guiding principles that will support your teaching journey. We share a multidimensional framework we call the *Culturally Responsive Mathematics Teaching Tool (CRMT2)* that reflects three interwoven strands: Knowledges and Identities; Rigor and Support; Power and Participation. There are three dimensions within each strand that can also be interwoven in a variety of ways to enhance your teaching practice. Each strand has its own chapter where we unpack each dimension with examples of instructional approaches, teaching stories, and try-it activities to illustrate what the dimensions look like in elementary classroom settings.

You will find a comprehensive view of the CRMT2 in the Appendix.

In addition, the framework is also formatted as a comprehensive and flexible teaching tool that can support your work in several ways, including unit and lesson design, peer observation and feedback, and curriculum adaptation. Each dimension has an essential question to help guide critical reflection on your teaching and a rubric scale with descriptors to help you envision what strong and centered CRMT can look like. We invite you to be curious and make connections to the fundamentals of CRMT—what resonates with you and what might stretch you as a culturally responsive mathematics teacher.

CHAPTER 1

....

CULTURALLY RESPONSIVE MATHEMATICS TEACHING: PURPOSE AND PRINCIPLES

Inside a fourth-grade classroom at a community charter school in Oakland, California, twenty students are in small groups around the room, with markers, pencils, papers, rulers, and other materials spread across their tables. They are fluently switching back and forth between Spanish and English as they take measurements and mark up maps. The children are working on a school-community math project: redesigning their school's future fifth-grade classroom building and play space. Each team is tasked to provide a list of recommendations along with visual maps to help the architect generate plans for the building remodel. A small group of students huddle around a map of their school's future fifth-grade outdoor space. The map is produced on grid paper and sits in front of another piece of paper with a four-column chart labeled "surface," "square yards," "fraction," and "purpose." As they talk, they gesture toward the paper, tracing lines on the map with pencils and markers. The energy level is high, as students encourage one another's ideas:

"So, what do we want to do with this space?"

"I think we should make a hangout zone, or wait maybe—"

"And a stage, but—"

"Maybe it'll be smaller than that," chimes in another student, tapping the map and gesturing with her hands, holding them about shoulder width apart.

"I don't want the stage area to be too big, but wait—oh, oh, oh! I got it! Maybe this part could be the stage, and this part

the audience?" a student asks, tracing a straight line through the group's largest open space on their map.

"But wait, we don't want to make it indoor, we really want it outdoor."

"Yeah," they all agree, "outdoor."

As the discussion continues, the three students fill the "surface" column with words like asphalt, grass, wooden stage, and trampoline. They get up from their seats and visit the class's paper representation of one square yard taped on the ground, asking one another, "Do we think 15 square yards is big enough?" They discuss how much space they need for each area, what kind of surface they would want (grass, asphalt, turf, wood, etc.), and calculate the fraction of the total area of the outdoor space using a previously calculated number of total square yards. (Adapted from Zavala & Singwi-Ferrono, 2018).

The preceding vignette is from the mathematics classroom of a fourth-grade teacher, Marianna, who was committed to social justice in her teaching. But, like many elementary teachers, Marianna struggled to realize those commitments within her traditional school mathematics curriculum. Dissatisfied with her students' engagement during math time, Marianna and author Maria—who was working as a teacher educator at the time—partnered up to plan math lessons together that would be different. They discussed ways to create a positive community of learners who were all valued. This meant creating lessons that positioned her students as *sense-makers* who are curious about their world, rather than *troublemakers* needing to be controlled—a stereotype that permeated schools with majority Brown children and families. Marianna wanted her students to feel ownership and agency over mathematics, bringing in their own mathematical ideas and lived experiences to mathematics learning. The school area redesign project was one way in which she could adjust mathematics instruction to meet those goals while students learned fourth-grade mathematics content standards. The students were excited, creative, and focused during the project. They originated mathematical ideas, used a variety of their own funds of knowledge, and constructed shared understandings of their community and mathematics, all while addressing a real issue in their community. These lessons, while a departure from how her traditional mathematics curriculum was structured, fostered mathematical freedom for her students and for Marianna as their teacher.

- Learn our definition and background of Culturally Responsive Mathematics Teaching (CRMT)

- Learn about three guiding principles to help steer our heart compasses as we engage in CRMT

OUR PURPOSE

In this book, we start with the heart, because our hearts beat life through our bodies and minds and evoke ideas of love, caring, and compassion. We see children as whole people; they are neither extensions of content or data nor solely adults in the making: They are living, breathing, thinking, emotional, and social—they are full human beings.

The sense of confidence and competence that children build in the elementary years is a resource for persistence in mathematics through middle and high school and also in their lives outside of the classroom. We argue that who students are in mathematical settings, their sense of positive mathematics identities, and their hearts are more important than standardized performance measures. To foster positive mathematics identities, we propose re-envisioning elementary mathematics classrooms to be places that embrace children as whole human beings, in light of who they are in their families and communities outside of school, and nurturing that humanity in and through mathematics instruction.

> *"We propose re-envisioning elementary mathematics classrooms to be places that embrace children as whole human beings, in light of who they are in their families and communities outside of school, and nurturing that humanity in and through mathematics instruction."*

WHAT IS CULTURALLY RESPONSIVE MATHEMATICS TEACHING?

We anchor this book in Culturally Responsive Mathematics Teaching (CRMT). CRMT is a comprehensive approach to teaching mathematics that offers conceptual and practical tools to support teachers from all backgrounds, identities, and political orientations to engage in the deep reflective work necessary to teach mathematics in ways that empower mathematical hearts in light of one's humanity—cultures, races, ethnicities, genders, sexualities, abilities, class positionings, communities, and languages. We recognize that this is a mouthful, so let us boil it down a bit. When we say, "culturally responsive mathematics teaching" (CRMT), we offer a definition that may differ from what you have previously heard.

CRMT involves a set of specific pedagogical knowledges, dispositions, and practices that privilege mathematics, mathematical thinking, cultural and linguistic funds of knowledge, and issues of power and social justice in mathematics education. CRMT interrogates and innovates mathematics instruction to be a transformative and humanizing experience for everyone.

The Roots of CRMT

This book is devoted to the practice of CRMT in classrooms, but we also believe that teaching is a scholarly activity. So, we start with the theoretical underpinnings (or roots) to acknowledge the traditions and research that CRMT draws on.

We situate CRMT in the intersection of three important constructs: Pedagogical Content Knowledge (PCK), Culturally Responsive Pedagogy (CRP), and Rehumanizing Mathematics (Aguirre & Zavala, 2013; Gutiérrez, 2018) (Figure 1.1). Understanding each of these constructs will help inform the practices of CRMT, which we will discuss in Chapter 2.

FIGURE 1.1 Roots of Culturally Responsive Mathematics Teaching

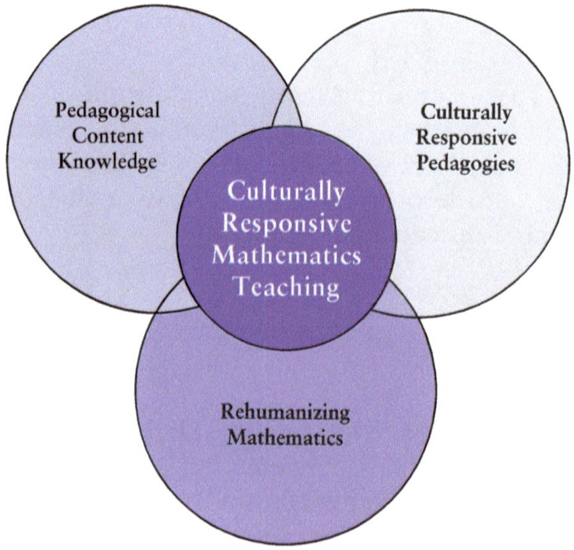

Pedagogical Content Knowledge

Pedagogical Content Knowledge (PCK) focuses on how to teach mathematics. It consists of general knowledge about teaching and its purposes; knowledge of student understandings, conceptions, and misunderstandings about mathematics; knowledge of curriculum; and knowledge of teaching strategies and representations related to teaching a specific topic such as partial products and arrays to teach multidigit multiplication (Grossman et al., 2005; Sowder, 2007). PCK is informed by theories of cognition with strong connections to constructivism (e.g., Piaget) and social constructivism (e.g., Vygotsky). In other words, students construct their own knowledge. They are sense-makers. Children also learn through social interaction and language. Peer collaboration and working with a "more experienced other" enhance cognition and development (Roth & Radford, 2010; Vygotsky, 1978).

However, PCK by itself is insufficient for effective mathematics teaching and learning, because it does not address issues of power that shape knowledge related to teaching mathematics, including what counts as the mathematical knowledge to be taught, who has access to mathematical knowledge and when, and what strategies support mathematics learning with children from various racial, cultural, linguistic, socioeconomic, gendered, neurodiverse, and (dis)ability backgrounds. Furthermore, the mathematical practices that occur outside the classroom in children's homes and communities are completely absent from the PCK framework for what teachers need to know about teaching mathematics (Civil, 2007). The other two constructs of CRMT address these issues.

Culturally Responsive Pedagogies

The term **Culturally Responsive Pedagogy** (CRP) was coined by education scholar Geneva Gay. It is grounded in liberatory pedagogy, ethnic studies, and multicultural education. According to Gay (2002),

> Culturally responsive pedagogy simultaneously develops, along with academic achievement, social consciousness and critique, cultural affirmation, competence, and exchange; community-building and personal connections; individual self-worth and abilities; and an ethic of caring. ... Culturally responsive teachers have unequivocal faith in the human dignity and intellectual capabilities of their students. They view learning as having intellectual, academic, personal, social, ethical, and political dimensions, all of which are developed in concert with one another. (pp. 43–44)

In CRP, teaching and learning are valuable and valued, caring and community-building, individual and cultural, as well as rigorous and socially conscious. All of these dimensions must be nurtured "in concert" with one another rather than ranked and prioritized. Students are sense-makers; they construct their knowledge through social interaction inside and outside the classroom (note the overlap here between CRP and some of the key aspects of PCK). Teachers must center their instruction around student knowledge and lived experiences to help bridge to new understandings. Teachers who might hold a more lock-step individual-oriented behavioral approach to learning cannot, by definition, hold a culturally responsive approach to teaching.

The political and social consciousness aspects of CRP stem from the liberatory pedagogy popularized by Brazilian education philosopher Paolo Freire through literacy campaigns with farm laborers in the 1970s (Freire, 1970/1993). Freire, who was exiled from his country for several years because of this work, helped teach people to read through co-constructing knowledge through problem-posing and praxis (i.e., learning as deep critical reflection and action, rather than passive reception of information). Freire described the oppressive education system as "banking education," in which learners are passive recipients of knowledge that is to remain unquestioned. The teacher "deposits" information into the head of the learner, maintaining the authority and ownership of knowledge. He contrasted banking education with liberatory education that emphasizes co-constructing knowledge through problem posing and inquiry. This approach to education situates the teacher and learner as co-learners investigating the world around them and changing it for the better. Knowledge is constantly being invented and reinvented through struggle, dialogue, innovation, and transformation. Freire (1970/1993) argued that to deny someone the opportunity to engage in inquiry is a dehumanizing and violent act. He stated,

> Any situation in which some individuals prevent others from engaging in the process of inquiry is one of violence. The means used are not important; to alienate human beings from their own decision-making is to change them into objects. (p. 66)

While Freire's focus was on literacy as a liberatory pathway, critical mathematics education scholars extend this argument to the mathematics education learning space through mathematical analysis of power relationships, fairness, and social/environmental/economic justice (Gutstein, 2006; Gutstein & Petersen, 2013). The liberatory connection among thought, power, learning, and transformation applies to mathematics as well. Author Julia made this connection explicit in a deeply personal piece about privileging equity and mathematics, writing, "to deny others the opportunity to engage in the process of mathematizing the world—to utilize mathematics to make meaning, connect to other forms of knowledge, and inform decisions—is an act of dehumanization" (Aguirre, 2009, p. 297). Grounded in liberatory pedagogy, CRP seeks to *rehumanize* such dehumanized learning experiences that foster inquiry, understanding, critique, and action for oneself and in relation to others and the world.

CRP also has roots in ethnic studies. During the height of the civil rights movement in the 1960s and 1970s, universities were experiencing student protests for a more balanced and just education. Historically marginalized students (people of color and women) attending universities argued that the required dominant canon of literature and history did not include multiple voices and perspectives. They argued for an academic space that reflected scholarship and stories grounded in the lives of the systematically marginalized, segregated, and oppressed. They demanded that the university curriculum be both a mirror and a lens for the increasingly diverse student body. They challenged the dominant, Western-centric, and patriarchal curriculum that perpetuated racist and sexist stereotypes. Thus, ethnic studies was born of the struggle for civil rights and human dignity to learn about histories that had been suppressed. In the early years, ethnic studies included African American Studies, Chicano studies, Asian American studies, Native American studies, and women's studies. Through this socially conscious and political struggle, ethnic studies broadened higher education and offered a counterbalance to the "classics" that often excluded women scholars and scholars of color.

At present, while what constitutes ethnic studies continues to evolve (see for example, new areas of Latinx, (dis)ability, and Queer studies), the fight for ethnic studies continues as state governments consider legislation to include ethnic studies as part of the general education requirement for K–12. This fight continues, in part, from analyses that came out following the struggle over Tucson Unified School District's ethnic studies program, which showed that Latinx students who took multiple Mexican American studies courses did better academically (e.g., state assessments and graduation rates) than a comparable cohort that did not. And yet, ethnic studies was labeled "divisive" and banned in a political battle that ignored the evidence of how it was actually good for Latinx students. The struggle in Tucson, which has repeated itself in other school districts in California and New York, is a case study in systemic racism involving equity-based and culturally responsive education (Cabrera et al., 2014). Ethnic studies approaches to mathematics education are being developed alongside the general development of K–12 ethnic studies curricula and contribute to how we conceptualize CRMT. Ethnic studies approaches to mathematics teaching are largely built off of the work in ethnomathematics (Furuto, 2014; Powell & Frankenstein, 1997) and critical mathematics (Frankenstein, 1983; Gutiérrez, 2010/2013; Gutstein, 2006). For example, in the Seattle Public Schools, systemic resources are being invested to create an ethnic studies framework for mathematics that includes components of agency, identity, power, oppression, and liberation to critically analyze the past and present actions with mathematics (see Gewertz, 2019).

While forward progress has been made, the mathematics education community still grapples with attending to the cultural, political, and social dimensions of mathematics teaching and learning. In teacher education contexts, CRP is often introduced as part of a required multicultural education course. However, the curriculum and instruction content examples offered in such courses rarely include mathematics, instead focusing on social studies and English/language arts content areas. This situation is further complicated when math methods courses or professional development offerings do not address the political, social, and cultural components of mathematics teaching and learning.

Rehumanizing Mathematics

Rehumanizing mathematics is the third key construct that informs our understanding of CRMT. Rehumanizing mathematics acknowledges the fact that mathematics is a human activity, one that has been done by people every day for thousands of years—mathematical activity that is free from "wealth, domination and compliance" (Gutiérrez, 2018). Ethnomathematics, which is informed by the fields of anthropology and mathematics, offers many examples of how mathematical activity is all around us—through geometric patterns in Navajo rug designs (Kirchner & Sarhangi, 2011); curves and spirals through cornrow braiding designs in hair (Eglash, 1999; Gilmer, n.d.); music rhythms and aerial patterns of cultural dance; family commerce and agricultural practices (Civil & Khan, 2001); cultural navigation and exploration (Furuto, 2014, 2015); architectural design; and concepts of time, space, and distance (Pinxten, 1997).

Thus, mathematics is deeply cultural—not culture-free. It is learned through language (verbal, nonverbal, symbolic, graphic) and is not language-free. Mathematics is embedded in how we measure and move in the world. Yet, mathematics introduced through schooling often removes this humanistic component by dismissing mathematical activity and knowledge outside of school as well as erasing both historical and current non-Western contributions to mathematical knowledge. School mathematics is scripted, disconnected, and fractured. Rehumanizing mathematics offers complexity, connectedness, and joy. People can be creative, intuitive, relational, and tactile as well as appreciate the logic and preciseness of mathematics. It is a more *freeing* and *broadened view* of mathematics—one that embodies wholeness (Gutiérrez, 2018).

Another important aspect of rehumanizing mathematics is addressing the roles of community and power. Gutiérrez (2018) argues,

> *rather than assuming a neutral response or failing to attend to power dynamics, rehumanizing mathematics recognizes that challenging the status quo will likely be met with great opposition from those with privilege and high status who benefit from the system remaining the same. It also seeks to highlight where power dynamics have played out in the history of mathematics and where mathematics might come to serve the people as opposed to vice versa … rehumanizing mathematics begins with the power of communities and assumes a relational view is important (recognizing oneself in others and others in oneself) so that we might better understand and live along side of one another and so*

that we might practice mathematics in ways that transform reality in emancipatory ways. (p. 4)

Imagine if this expansive, transformative, rehumanized view of mathematics was normalized as school mathematics. How might that affect the ways students experience mathematics and teachers teach mathematics? CRMT embraces this view of rehumanizing mathematics.

OUR GUIDING PRINCIPLES

If our purpose is to move toward a rehumanizing, liberatory mathematics education through CRMT, there are three fundamental principles that can help anchor our decisions and situate ourselves when faced with dilemmas or tensions in our practice.

1. **Acknowledge that our current system of mathematics education is inequitable and oppressive.**

    Our current mathematics education system privileges people from particular classed, raced, languaged, gendered, abled, and cultured backgrounds and oppresses those who do not belong to those privileged groups. This starts with what counts as mathematics (e.g., "math started with the Greeks"), who authors the mathematics (textbook companies and state assessment designers), and what mathematical competence gets valued in the classroom (often, being correct and fast when solving procedural problems). This can make us feel like mathematical competence is innate or is historically predetermined by prescriptive curriculum and standards with limited options for teachers or students to be active authors of mathematics experienced in the classroom.

    For many children from nondominant groups, mathematics is a tool of violence and segregation (Martin, 2012, 2015). It is used to rank and sort children based on perceived cognitive deficits identified through a lens and legacy of racial hierarchies reflected in standardized testing. These tests include "readiness" tests used before entering kindergarten, state assessment tests that start in third grade, or IQ tests used to determine placement in coveted "highly capable" classrooms (Berry et al., 2014; Ellis, 2008). Research consistently shows Black, Brown, and Indigenous children, multilingual learners (MLL), and children from working-class communities are overrepresented in special education and remedial services and underrepresented in "gifted" programs and advanced placement courses. Forty years of research on tracking (also known as "ability grouping" in elementary settings) has shown that ranking and sorting children using mathematics negatively impacts children's educational and life trajectories (Boaler, 2002;

Boaler & Selling, 2017; Oakes, 1985/2005). These tracks are racialized, classed, and gendered—resulting in vast differences in educational persistence and experience, especially in STEM fields, and overall underperformance of students who do not have the financial resources to provide extensive out-of-school supplemental instruction by private tutors or corporations such as Sylvan or Kumon.

Furthermore, there is a systemic misunderstanding that mathematics is language-free and universal, meaning that mathematics knowledge requires less language than other subjects—"numbers are just numbers." This view of mathematics is patently false but often used when talking about supporting MLL. Language matters in learning and communicating mathematics (de Araujo et al., 2018; Moschkovich, 1999; Turner & Celedón-Pattichis, 2011). Even symbolic notation varies across the globe. There are cultural differences in representing numbers and algorithms. For example, in many Latin American countries, $59 : 8 = 7 + 3 : 8$ is a way to symbolically represent the equivalent fractions $\frac{59}{8} = 7\frac{3}{8}$ (Perkins & Flores, 2002).

Unfortunately, this misunderstanding about mathematics as universal and language-free shapes our assessment systems. For example, in the state of Washington, newly arrived immigrant children who are just learning English must take the computer-based state mathematics assessment in English within the first year of their arrival. There is a three-year grace period for the literacy state assessment. This means that children who may come from war-torn countries and may not have been in school for long periods of time must take a high-stakes mathematics assessment in a new language using unfamiliar technology. This can be a traumatizing and violent experience for these children and their families, this time with mathematics instead of bullets or bombs.

Acknowledging that our current system is inequitable and oppressive reminds us that entrenched patterns of educational outcomes can only be disrupted if we see them and work to change them. Marianna recognized that the traditional mathematics taught in her classrooms was disconnected from students' lives. This is what Marianna wanted to change. She realized that making mathematics relevant to her students could be a make-or-break motivator, especially for those already feeling pushed out or left behind by schooling even at the tender age of nine. Her actions disrupted what was traditionally taught and opened up new and more joyful opportunities for her students to learn.

An important starting point for this principle is to repeatedly remind ourselves to ask the questions, *Who was school mathematics originally intended for, and to whose benefit? Is school mathematics working for all my students? If not, what can I change?*

2. **Take actions that center students and their families inside and outside the mathematics classroom.**

This principle probably speaks the most to us teachers. We take action. We are change agents. In mathematics education, in particular, we can take actions to challenge who has access to rigorous mathematics and expand whose mathematics is legitimized in classrooms. Our actions must work to center cultural, linguistic, and other funds of knowledge throughout the math classroom. When we say center, we mean center. That is, families and students cannot be an afterthought to an already-planned lesson. Rather, families and communities outside of school must take a top priority. This means to recognize that children engage with mathematics prior to and beyond schooling. Teachers can take action and leverage the mathematical knowledge in families and communities as mathematical resources for lessons. In the building redesign vignette, Marianna's students were familiar with measurement concepts and tools. They wondered about the concept of area in relation to play activities. And they applied understandings of fractions to figure out the area of subunits of the whole space they had to work with. They drew upon both their outside and inside school knowledge to think about space and how to measure it.

However, we also recognize that it is challenging to center mathematics instruction on families when a teacher is under pressure to use the textbook "with fidelity" or "teach to the test." That is why we have to have clear principles. We must remind ourselves of what is at stake here and help us decide at times to deviate from the text in creative ways that still engage students in meaningful mathematics, like Marianna did. By centering students, families, and communities, we set the stage for mathematical empowerment. Therefore, our actions as mathematics teachers must include attention to important questions like, *Who is really empowered in my classroom? How can my instruction facilitate empowerment among my students and their families?*

3. **Be accountable to ourselves, our children, our families, and our communities.**

Mathematics teachers are identity workers (Gutiérrez, 2013). This means that whether we know it or not, we shape how children experience mathematics, feel about mathematics, and see themselves as mathematics learners and doers. We hold power to affirm or dismiss all of the identities and lived experiences children bring with them to the classroom. Beyond test scores (again, who benefits?), we need methods to hold ourselves accountable to the change we would like to see in our schools.

Holding oneself accountable is a way to move toward disrupting our broken system. It requires teachers to understand their role as identity workers and to be genuine learners with children and families about mathematics, student thinking, culture, language, and other aspects of responsive pedagogy. Learning alongside, about, and with children's communities may require initial discomfort or disequilibrium, as teachers learn more about themselves and the communities of families that may be very different from themselves. Discomfort is not necessarily a bad thing. We can get curious about our discomfort. We can hold ourselves accountable while also being compassionate with ourselves as we try new things.

Accountability should not happen in isolation. Culturally responsive mathematics teaching emphasizes community and collaboration alongside individual agency to make decisions. Teachers need to work together on a regular basis, to lift each other up in their teaching practice, to ask essential questions about their teaching practice, as well as to hold each other accountable for the honor of teaching children. Marianna collaborated with author Maria to hold herself accountable and try new things that connected with her commitment to social justice. She expanded the mathematics curriculum beyond the text by creating a set of math lessons driven by an important and very real community-based situation that mattered to her students. We can hold ourselves accountable through the *mirror test* (Gutiérrez, 2016), in which we look at ourselves and ask, *Am I doing what I said I wanted to do in education when I set out to be in this profession? And if I'm not, what am I going to do about that?*

ESSENTIAL QUESTIONS FOR CRMT

As a preview to Chapter 2, we want to highlight essential questions teachers designing math instruction through the lens of CRMT ask themselves. These are the kinds of questions that guided Marianna and Maria to design their mathematics lessons for the play-space redesign project in the opening vignette. How does my lesson …

- **help** students connect mathematics with relevant/authentic issues or situations in their lives?

- **support** creativity, broaden what counts as mathematical knowledge, and affirm positive mathematical identities for all students?

- **create** opportunities to elicit, express, and build on student mathematical thinking in multiple ways?

- **enable** all my students to closely explore and analyze math concept(s), procedure(s), and problem-solving/reasoning strategies?

- **maintain** high rigor with high support for all students?

- **make space** for MLL to be central participants in mathematics activities?

- **distribute** mathematics authority and make space for multiple forms of knowledge and communication?

- **disrupt** status differences, entrenched stereotypes, and inequitable power relationships present in all mathematics classrooms?

- **encourage** student use of mathematics to analyze, critique, and address power relationships and injustice in their lives (economic, social, environmental, legal, political, patriarchal)?

These essential questions guide potentially new and innovative practices. They reflect ways of designing mathematics learning experiences that Marianna and her students found freeing. Such practices disrupt the current oppressive systems and, if done regularly and collaboratively, could permanently transform how children experience learning mathematics for the better. In the chapters that follow, we will introduce a tool that can help you make answering these questions part of your planning and reflection.

TOWARD HUMANIZING AND JUST MATHEMATICS LEARNING EXPERIENCES

Our main purpose for writing this book is to move us, as educators, toward a more humanizing and just experience with mathematics, especially in school. Collectively, we respect our roles as teachers who deeply care for the children and families we serve. We believe that culturally responsive mathematics teaching in elementary school is a promising pathway to make this happen.

DISCUSSION QUESTIONS

- Reflect on the reality: When was the last time you taught mathematics in a way that was freeing for you and your students? What made it that way?

- Expand the vision: Imagine if this expansive, transformative, rehumanized view of mathematics was normalized as school mathematics. How might that affect the ways students experience mathematics and teachers teach mathematics?

CHAPTER 2

· ·

UNPACKING CULTURALLY RESPONSIVE MATHEMATICS TEACHING: RETOOLING YOUR PEDAGOGY

Simon is nine years old. Every Saturday he accompanies his father to work restocking vending machines in various buildings in the downtown area. While his father restocks, Simon removes the money from the machine. There are always different kinds of coins: quarters, nickels, and dimes. Bills too, but mostly coins. Simon rolls the coins in special paper marked $10 quarters, $5 dimes, and $2 nickels. Bills are put into an envelope. Simon writes down the total amounts from each machine in a bank book. When the day is complete, they head to the bank to deposit the money.

This scenario illustrates the family work and economic practices of a nine-year-old boy named Simon. Simon helps his father to restock vending machines and has a specific responsibility to collect, sort, and package coins and bills. He must also determine and record total amounts.

PAUSE AND REFLECT

- How could a teacher use this information to teach children mathematics?
- How could a teacher use this information to teach mathematics to *this child* in particular?

Perhaps you thought about posing questions to children related to money, such as finding total amounts based on vending machine data. Perhaps you thought about exploring place value with authentic amounts written in the bank book. Perhaps you thought that Simon might especially shine if given math situation problems that included money or currency exchange contexts. While money is a common context used in mathematics textbooks for application problems, we would like you to think beyond money. Instead, we invite you to think about other mathematical concepts or domains that might be explored with this situation. Some examples for student exploration might include the following:

- Optimize transportation routes using maps with locations of the vending machine and traffic information

- Predict how many snacks are in the vending machines over a period of time

- Identify the proportion of sweet to nonsweet snacks that are in the vending machines

- Determine how changes to the position of two snacks, three snacks, or four snacks in the vending machine affect the settings of the vending machine

Solving any of these math situations would engage students in high cognitive demand activities. Depending on how the activity is structured, students might be working in groups, sharing their math thinking and connecting to different kinds of knowledge and experiences to answer these questions.

Now, for some additional information. Suppose you found out that Simon was a Black boy of Haitian descent and bilingual (Haitian-Creole/English). Would your ideas for teaching children mathematics change? Would your ideas for teaching math to Simon, in particular, change?

PAUSE AND REFLECT

- Does this new information about this child's racial, ethnic, and language background make any difference? Why?

As culturally responsive mathematics teachers, we are identity workers (Gutiérrez, 2013). A child's whole self needs to be honored through our teaching. Our mathematics curriculum is a mirror and a lens. So, to move beyond a surface-level connection to the child and the mathematics, we would want to ask ourselves a number of questions, such as the following:

> *"As culturally responsive mathematics teachers, we are identity workers (Gutiérrez, 2013)."*

- How might we leverage the mathematical, cultural, and community-based funds of knowledge this child brings to his classroom?

- How do we affirm this child's mathematics identity?

- In what ways can his lived experiences be leveraged to deepen his own and other children's mathematical understandings?

- What math ideas can be elicited and built upon?

- Being bilingual, how might this child understand and communicate mathematical ideas to his peers and to the class? How are those contributions valued?

- How might social status hierarchies that historically benefit white, monolingual English-speaking children be addressed to welcome and affirm the brilliance of this Black child?

Responses to these questions are necessary and complex as we think about customizing our mathematics instruction to reach and teach each and every child we work with in the classroom. What would mathematics teaching look like if we designed our instruction with these questions and this child in mind? While it may seem overwhelming, what are the consequences if we don't?

BIG IDEAS IN THIS CHAPTER

- An overview of an instructional planning tool for Culturally Responsive Math Teaching (CRMT)

- A summary of various ways to use the tool to support your CRMT journey

INTRODUCING THE CULTURALLY RESPONSIVE MATHEMATICS TEACHING TOOL (CRMT2)

We have developed a professional tool to streamline the process, which we call the Culturally Responsive Mathematics Teaching Tool (CRMT2). This tool provides a multidimensional view of mathematics teaching that attends to the strengths and complexities of children, families, interactions, and systems that can shape math teaching and learning in your classroom. In particular, the tool will help you attend to teaching mathematics with strong, deep, and meaningful connections to students and their communities.

In our previous work, we have found that reflection tools are quite helpful in providing guidance to analyze instruction both individually and with colleagues in professional learning communities (Aguirre & Zavala, 2013; Aguirre, Zavala & Katanyoutanant, 2012). Such tools must be versatile and useful for many purposes including unit planning or lesson design, periodic tune-ups of one's own teaching, peer feedback, and other forms of self-reflection. In addition, such tools must also capture the multidimensionality of CRMT in an accessible and familiar way.

We call the CRMT2 a design tool because we view teaching as a design process: creative and dynamic work through which teachers make a deep impact on the world through their work with children and families. While designers engage in cycles of development, prototyping, and testing, we teachers do the same but call it planning, teaching, and reflection.

The Structure of the Tool

The CRMT2 focuses on three strands: (1) Knowledges and Identities; (2) Rigor and Support; and (3) Power and Participation. Each strand has a unique set of three dimensions, each of which has an essential question and a set of rubric descriptors. Figure 2.1 summarizes the three main strands and their respective dimensions. You will notice that the dimensions are grounded in the definition, principles, and practices described in Chapter 1. We associate each strand with a particular color, the traditional colors of New Orleans's Mardi Gras celebration: green (faith), gold (power), and purple (justice).[1] To see the strands in full color, please look at the inside cover or go to corwin.com/books/mathematical-heart-277718. In addition, the tool is set up in a rubric style with a rating scale from 1 to 5. The rating scale depicts a progression of strength and centeredness in teaching practice with a rating of 1 representing fragile and on the margins of teaching practice and a rating of 5 reflecting strong and fully centered in teaching practice. The rubric for each strand and related dimensions will be discussed in detail in subsequent chapters. Here we briefly describe the three strands with their accompanying dimensions and essential questions. Then, we discuss how you can use the tool in a variety of ways to promote critical reflection and professional feedback to improve practice. Ideally, we want this tool to be useful for enriching the math learning environment and experience.

FIGURE 2.1 CRMT2 Overview

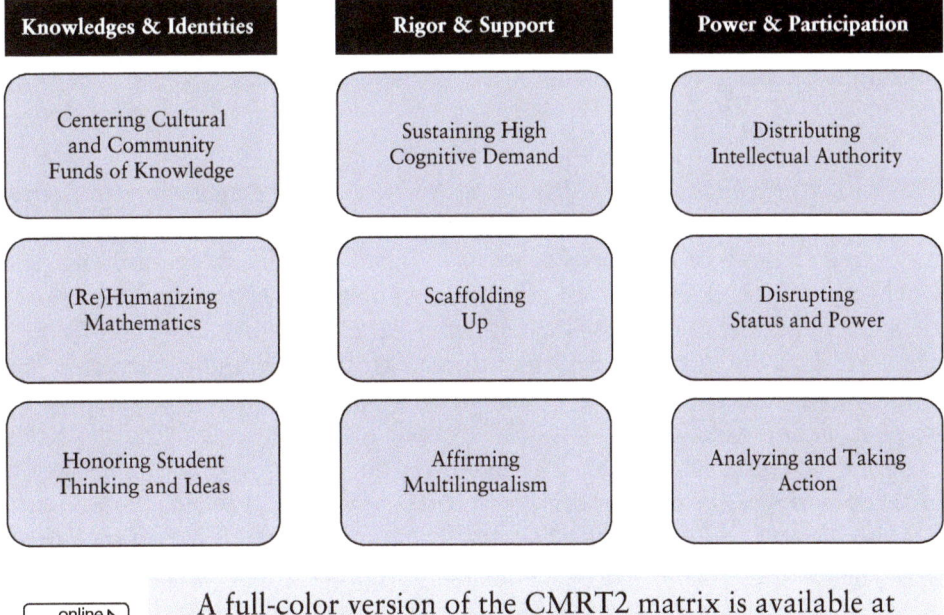

Knowledges & Identities	Rigor & Support	Power & Participation
Centering Cultural and Community Funds of Knowledge	Sustaining High Cognitive Demand	Distributing Intellectual Authority
(Re)Humanizing Mathematics	Scaffolding Up	Disrupting Status and Power
Honoring Student Thinking and Ideas	Affirming Multilingualism	Analyzing and Taking Action

online resources

A full-color version of the CMRT2 matrix is available at
corwin.com/books/mathematical-heart-277718

---

[1] https://www.visitjeffersonparish.com/events/festivals/mardi-gras/mardi-gras-history/#:~:text=The%20traditional%20colors%20of%20Mardi,gave%20the%20colors%20their%20meanings.

Strand 1: Knowledges and Identities

The first strand of the CRMT2 focuses on **Knowledges and Identities**. This strand is expanded on in Chapter 3. Both knowledges and identities impact how we learn and see ourselves as learners. There are three dimensions in this strand: *Centering Cultural and Community Funds of Knowledge, (Re)Humanizing Mathematics*, and *Honoring Student Thinking and Ideas*. These dimensions reflect fundamental concepts about what matters for CRMT. The color for this strand is green. The eminent scholar Geneva Gay (2000), who coined the term "Culturally Responsive Teaching," asserts that culturally responsive teachers must have "*unequivocal faith* in the human dignity and intellectual capabilities of their students (emphasis added)." The color green symbolizes this unwavering faith that students can and will learn mathematics in meaningful and deep ways.

Centering Cultural and Community Funds of Knowledge

We intentionally position cultural/community funds of knowledge as the first dimension. We do this to explicitly center the work on children and families. We acknowledge that all children come to classrooms with knowledge and experiences nurtured through family and community interaction. These knowledge bases, including math knowledge, can be leveraged to support meaningful learning (Civil, 2007; Turner et al., 2012). The essential question for this dimension is, *How does my lesson help students connect mathematics with relevant/authentic issues or situations in their lives?* We invite you to design learning experiences that leverage children's funds of knowledge with mathematical concepts. This supports an intentionality to make authentic connections to mathematics and bridge inside and outside of school learning.

(Re)Humanizing Mathematics

This dimension focuses on cultivating children's innate creativity and curiosity. Many children experience mathematics as a set of disconnected procedures and skills to be memorized. Joy is stripped from learning mathematics. Curiosity is set aside in favor of efficiency. And, in some cases, children experience dehumanizing processes when deficit labels such as "low" and "slow" are used. This dimension is designed to counter that dominant narrative. We embrace Rochelle Gutiérrez's statement that "*mathematics needs people as much as people need mathematics.*" We need to rehumanize the mathematics experience (Gutiérrez, 2018). The essential question for this dimension is, *How does my lesson support creativity, broaden what counts as mathematical knowledge, and affirm positive mathematical identities for all students?* The focus on identity is crucial here as children develop beliefs about themselves as doers of mathematics and are positioned by others (both adults and children) as doers of mathematics (Aguirre et al., 2013; Martin, 2000). We must reclaim and cultivate the joy, creativity, and curiosity of learning mathematics. And as both essential workers and identity workers, teachers are well-suited to lead this effort.

Honoring Student Thinking and Ideas

This dimension focuses on leveraging the thinking strategies and ideas children bring to and develop while learning mathematics. Grounded in a deep research base, learning mathematics is not a linear process (Carpenter et al., 1999/2014; Lampert, 2001). Children bring diverse ideas to solving mathematics. Sometimes these ideas are grounded in their lived experiences, and other times they are the result of cognitive dissonance as students make sense of new situations and try to connect to what they already know. You can do this best through social interaction and communication. The essential question for this dimension is, *How does my lesson create opportunities to elicit, express, and build on student mathematical thinking in multiple ways (e.g., through gestures, pictures, words)?* We encourage you to create learning environments that support and honor the diverse ways children may think about and express their mathematical ideas. It is key that students' ideas are elicited and valued in the mathematics classroom.

Revisiting the Vignette

In the case of Simon, we know he engages in a frequent mathematical activity having to do with his family's business. To leverage his funds of knowledge, we may design a task related to ordering supplies for a vending machine business or make a plan to service the machines in a specific time period. Students might begin with a mathematizing-the-world routine in which they are shown a picture or video about vending machines and asked what they notice, what they wonder, and what math questions they might pose (Aguirre et al., 2022; Turner et al., 2018). Students would be invited to pose problems about the situation. Simon can share his experience with vending machines and helping his dad. To design a specific task, Simon's dad might provide essential information about product (least/most popular items, price, size), service timelines, and locations of machines. A teacher might use these resources to help students identify important quantities necessary to answer specific questions students pose. Such activities help you affirm your students' identities and diverse ideas, including Simon. Furthermore, centering Simon's experience servicing vending machines with his dad as an authentic mathematics space positions Simon and his family as *intellectual resources* for mathematics teaching and learning in the classroom.

Strand 2: Rigor and Support

The second strand of CRMT2 focuses on **Rigor and Support**. This strand is expanded in Chapter 4. Rigor and support focus on instructional practices designed to open up access to math content and nurture mathematical proficiency including conceptual understanding, problem-solving/reasoning, and procedural fluency (National Research Council, 2001). There are three dimensions in this strand: *Sustaining High Cognitive Demand, Scaffolding Up*, and *Affirming Multilingualism*. These dimensions focus on creating mathematical learning experiences that embody high expectations with high support. The color of this strand is gold, which symbolizes power. In this context we are talking about *mathematical empowerment* developed through opportunities to engage in high cognitive demand activities, with supports that provide access to and advancement in mathematics, and in ways that leverage linguistic strengths and cognitive skills many multilingual learners (MLL) bring to the classroom (Celedón-Pattichis & Ramirez, 2012; Chval et al., 2021; Civil & Turner, 2014).

Sustaining High Cognitive Demand

This dimension attends to the level of complex thinking experienced by students in a math lesson (Smith & Stein, 1998). High levels of cognitive demand ensure students are engaged in complex problem-solving and making connections between concepts and procedures—knowing the why. Low levels of cognitive demand emphasize rote memorization and mastery of procedures with limited conceptual connections. There is a mathematical myth that suggests students need to learn by rote facts and procedures before they can engage with more complex rigorous tasks. This is not true. Students can and have engaged with complex mathematics that helps them learn facts and procedures with understanding (Carpenter et al., 1999/2014; Smith & Stein, 2018). The essential question for this dimension is, *How does my lesson enable all my students to closely explore and analyze math concepts(s), procedure(s), and problem-solving/reasoning strategies?* This dimension helps you design math lessons that invite curiosity, analysis, and problem-solving. Students' critical thinking skills become stronger when they engage in tasks that are novel and invite them to make connections to multiple mathematical knowledge bases including their invented math strategies and their funds of knowledge. The more you practice engaging complex tasks, the better you become.

Scaffolding Up

This dimension focuses on the scaffolds and supports needed to open up access and sustain student engagement in high cognitive demand activities. We invite you to design supports that build on students' strengths throughout the entire lesson. The essential question for this dimension is, *How does my lesson maintain high rigor with high support for all students? Scaffolding Up* includes deliberately planning supports that focus on individual student needs or specific needs of subgroups of students. In this dimension, you pay careful attention to preplanned supports and in-the-moment interactions that maximize student engagement during complex problem-solving.

Affirming Multilingualism

This dimension centers attention on the strengths and contributions of MLL. It emphasizes different strategies that build multilanguage skills including translanguaging, academic language development, and multimodel communication (Chval et al., 2021; Maldonado Rodriguez et al., 2020). The essential question for this dimension is, *How does my lesson make space for MLL to be central participants in mathematics activities?* This dimension helps you find multiple ways to affirm and grow your multilingual students' language and math learning. Whether you are monolingual, bilingual, multilingual, or somewhere in between, you can always *Affirm Multilingualism*.

Revisiting the Vignette

Cheche Konnen means "search for knowledge" in Haitian-Creole. This phrase invites students to be curious about the world, to investigate, analyze, and make decisions about data to solve real-world problems (Warren & Rosebery, 1991). In a culturally responsive mathematics classroom, Simon, a multilingual learner and Haitian-Creole speaker, is encouraged by this goal, which is reflected in his home language and English. He and his classmates routinely engage with complex problem-solving tasks that invite multiple perspectives and strategies. The language demands (reading, writing, speaking, and listening) are high in the classroom. In a language-rich math classroom, you can position MLL as intellectual resources by providing intentional support. For example, you can use sentence starters and graphic organizers to help students communicate their ideas to a broader multilingual audience. In addition, you can ensure MLL and their families have access to internet-based bilingual math resources to help navigate math curriculum when needed.

Strand 3: Power and Participation

The final strand of the CRMT2 tool is **Power and Participation**. This strand is expanded in Chapter 5. It focuses on instructional practices that broaden participation and the distribution of knowledge—centering students and minimizing status issues in the classroom. There are three dimensions in this strand: *Distributing Intellectual Authority*; *Disrupting Status and Power*; and *Analyzing and Taking Action*. The color of this strand is purple, which symbolizes justice. Justice-focused teaching disrupts status and power inequities that limit students' intellectual agency, their community-building, and their ability to make mathematical connections to analyze and solve meaningful problems impacting their lives (Gutstein, 2006).

Distributing Intellectual Authority

This dimension focuses on broadening the sources of knowledge present in your mathematics classroom. Instructional practice in this dimension centers math authority with students rather than with the teacher or textbook only. The essential question for this dimension is, *How does my lesson distribute mathematics authority and make space for multiple forms of knowledge and communication?* Here, we invite you to examine how you involve students in math learning. This includes strategies that cultivate space for diverse thinking and ways of communicating.

Disrupting Status and Power

This dimension aims to dismantle inequitable power relationships and status in mathematics classrooms that do harm to students' math identities, learning, and social interactions. This dimension promotes instructional strategies that affirm collaboration rather than competition; build on student's mathematical, cultural, and linguistic strengths; and create counternarratives to dominant racial and gender stereotypes that have long-term negative effects on student identities as math learners (Featherstone et al., 2011; Gholson & Robinson, 2019; Horn, 2012; Martin et al., 2010). The essential question for this dimension is, ***How does my lesson disrupt status differences, entrenched stereotypes, and inequitable power relationships present in all mathematics classrooms?*** This dimension encourages you to find different ways of minimizing status differences and maximizing involvement of students from historically segregated and marginalized communities.

Analyzing and Taking Action

This dimension emphasizes critical thinking by engaging students in using mathematics as an analytical tool to understand and address issues of power, access, representation, justice, and fairness in their communities. The essential question for this dimension is, ***How does my lesson support student use of mathematics to analyze, critique, and address power relationships and injustice in their lives (economic, social, environmental, legal, political, patriarchal)?*** This dimension helps you facilitate opportunities for students to be change-agents and take action to address inequities and injustices impacting their families and communities.

Revisiting the Vignette

In relation to the strand of power and participation, imagine Simon's math class engaging in a complex problem-solving task about access to healthy snacks via vending machines. Inspired by a youth activist who created a petition to remove junk food from the vending machine at a local community center, the students prepare to assess the quality and access of snacks in the vending machines in school. Intellectual authority is widely distributed as students offer different ways to measure the "healthiness" of the snack offerings and create a plan to order healthy snacks. The task also invites students to investigate environmental justice issues like food deserts often impacting working class communities of color—mathematizing community justice issues and taking action to assess and address this situation in their school and local communities.

HOW TO USE THE CRMT2

You may be wondering, how do I use such a lesson analysis tool as part of my regular classroom practice? Here we offer some explicit ideas that highlight the flexibility of the tool, and we'll explain each dimension of the tool more fully in the coming chapters so that you can get a better sense of what it looks like in practice. We also want to encourage you to engage creatively with it based on your own context.

CRMT2 to Analyze Math Teaching Practices

One key way we recommend using the tool is as a holistic source of reflection. The tool can also help you look critically at your own patterns in teaching practices, as described below, through self-reflection or observations with trusted colleagues.

Self-Reflection

In our previous work, teachers have used the CRMT2 to analyze the lesson plan of a recently taught math lesson, and you can do the same. You can self-assess within each dimension based on what was in your lesson plan and what happened during the lesson. You can rate your own teaching within each dimension to determine strengths and strands for future growth, because every lesson has strengths and room for adaptation or expansion. It is an opportunity to rethink and strengthen how you might teach the lesson next time or what you might do the next day. Research tells us that strong teachers deeply and consistently reflect on their practice (Zeichner & Liston, 2014). This is a tool that will help with that process.

Peer/Team Observation and Feedback

You can also use the CRMT2 for peer observation and feedback. You can self-select specific strands or dimensions you would like feedback on from a trusted colleague or coach. We recommend that you focus on at least three dimensions or one strand to start with. In addition, you might invite your colleagues or a coach to provide comprehensive feedback across all three strands. Either way, this maximizes opportunities for you to get meaningful and comprehensive feedback about CRMT from a colleague or coach in a supportive professional learning community.

CRMT2 to Design, Analyze, and Adapt Curriculum

Another way you can use the tool is for analyzing and adapting curricular materials, both those supplied as tools for your own mathematics teaching (for example, published curriculum) and those you create from scratch. The tool can be a useful anchor in a design process, providing prompts (essential questions) to guide your design, or you can use it to analyze and adapt existing lessons within the curriculum prior to instruction.

Checkpoints in a Curricular Design Process

Inspiration to make up our own mathematics units or lessons hits us all the time—let's be honest, when we love to teach mathematics, we look for inspiration everywhere. You can use the CRMT2 to guide that process or as a checkpoint as you start to flesh out a particular task or lesson plan for your students. The essential questions can help you decide, *Am I ensuring that various threads of CRMT are present?* before you teach a lesson you design from scratch. Similarly, it can provide a starting point for reflection once you have taught the lesson you designed.

Analysis and Adaptation of Curricular Materials

A key assumption we hold as culturally responsive mathematics teachers is that even curricular materials tailored to one's own school district need to be read, analyzed, and adapted by teachers to be responsive to their particular classroom of students. Certainly, not every lesson can get scrutinized in the same way. Sometimes, what you have time for is a quick reading over what the lesson is for today and the time to jot down some very minor changes. But other times, you may want to dive deep into your own processes of analysis, adaptation, instruction, and reflection. The tool can be used in this way, to support your analysis and adaptation. First, use the tool to review a lesson you plan to teach by evaluating what is written there with your students in mind. Evaluate the lesson for each category. For categories that had lower scores, consider how you will adapt the lesson to make this stronger. Put together the whole lesson with your adaptations, and then return to the tool to engage in self-reflection on how the lesson went afterward.

These kinds of curricular designs and adaptations don't have to be done alone. We encourage a creative process of co-constructing mathematics lessons, with children or colleagues or parents, as well as collaboration around critiques of published curricula. The more teachers working on analysis of and adaptations of curriculum who all teach in the same school or community the better!

Who Is Best Suited to Analyze for CRMT? Those Closest to the Context

The final point we will underscore is that the CRMT2 is best used by people who know the families and communities as well as the context in which the lessons will be taught. It may seem tempting to hand the work of curriculum analysis over to someone who could score it for culturally responsive teaching, but if you look at the essential questions on the tool, they imply a level of responsiveness that doesn't come from strangers. Those with proximity to the classroom, the students themselves, and those in their communities are best positioned to engage in the analysis and propose adaptations.

ACTIVITY 1: Starting Point

To help you get started with this work, we invite you to take the CRMT Confidence Survey. This will give you a starting point to reflect on your beliefs and practices related to CRMT. What resonates with you? Right now, what aspects do you feel most confident/least confident in? Think about why that might be the case and how you might build on those strengths and stretch yourself. After you read the book and work with the tool, we invite you to take the survey again and analyze your strands of growth.

The CRMT Confidence Survey

1. **Rate your confidence level** for each of the following statements. We invite you to reflect on what you notice and wonder about your self-ratings as you get started with the text.

	NOT CONFIDENT	A LITTLE CONFIDENT	SOMEWHAT CONFIDENT	VERY CONFIDENT	EXTREMELY CONFIDENT
Deepening my own mathematical learning	1	2	3	4	5
Strengthening my own math-teaching identity	1	2	3	4	5
Understanding math content	1	2	3	4	5
Designing math lessons that center community/cultural funds of knowledge	1	2	3	4	5
Partnering with families as intellectual resources for math instruction	1	2	3	4	5
Strengthening student math identities	1	2	3	4	5
Eliciting student math thinking in multiple ways	1	2	3	4	5
Analyzing student math thinking in multiple ways	1	2	3	4	5
Analyzing the cognitive demand of a math lesson	1	2	3	4	5
Analyzing language demands of a math lesson	1	2	3	4	5
Designing math lessons that promote student exploration and analysis	1	2	3	4	5
Differentiating instruction for students with learning differences	1	2	3	4	5
Designing supports for students with varied mathematical strengths	1	2	3	4	5
Affirming multilingualism to support math learning	1	2	3	4	5
Increasing student participation and engagement in math learning	1	2	3	4	5
Sharing math authority among students and teacher	1	2	3	4	5
Disrupting status differences in the math classroom	1	2	3	4	5
Addressing entrenched stereotypes in math	1	2	3	4	5
Connecting to social justice/critical knowledge in a math lesson	1	2	3	4	5
Making connections among math content, student math thinking, and community funds of knowledge in a math lesson	1	2	3	4	5

2. **Analyze your results.** What goals for professional and personal growth will you set as a Culturally Responsive Teacher of Mathematics?

To strengthen my culturally responsive mathematics teaching, I will . . .

To support my students' mathematics learning, I will . . .

LOOKING AHEAD

In the next few chapters, we will take a deeper look at each strand and the affiliated dimensions. We provide detailed descriptions, classroom vignettes, and teaching activities to help you visualize and enact these practices in the classroom. In Part 2, we include four chapters that explore CRMT in a variety of elementary classrooms, including primary (Grades K–2) and intermediate (Grades 3–5) as well as bilingual and special education settings. Our guest authors provide various examples of how the CRMT2 helped them plan, enact, and reflect on their own teaching of mathematics. We invite you to continue this journey to cultivate mathematical hearts through CRMT.

DISCUSSION QUESTIONS

- How does your math teaching connect to faith, em(power)ment, and justice?

- What thoughts or feelings were evoked when you completed the CRMT confidence survey?

- What CRMT strands (Knowledges and Identities; Rigor and Support; Power and Participation) resonate with you and stretch your thinking?

CHAPTER 3

·······························

KNOWLEDGES AND IDENTITIES

Knowledges & Identities	Rigor & Support	Power & Participation
Centering Cultural and Community Funds of Knowledge	Sustaining High Cognitive Demand	Distributing Intellectual Authority
(Re)Humanizing Mathematics	Scaffolding Up	Disrupting Status and Power
Honoring Student Thinking and Ideas	Affirming Multilingualism	Analyzing and Taking Action

Alicia is a third-grade teacher in her second year at a public school in the diverse Mission District of San Francisco. It is October of the school year. She revels for a moment in the quiet settling on the school, now that the children are off in after-school activities: maybe skipping home with their parents, maybe stopping at the panaderia on the corner, or the city library across the street, as they walk to the transit stops or walk home. She smiles slightly, remembering Arthur, one of her quietest students, grinning as he handed her a fortune cookie he had saved from his father's restaurant the night before. "Para ti" he had said, and she was so proud to hear him voluntarily speak Spanish, a language many children spoke in the classroom, a language he was learning alongside his other two, Mandarin and English. She momentarily glimpses over at Sandra's desk. At times, Sandra's mom took her to work with her, cleaning houses and babysitting, if she couldn't find someone to pick her up after school. Alicia puts a reminder into her phone to call Sandra's *mom and then turns her attention to the task at hand: preparing to launch a unit on multiplication that she taught one time before the prior year.*

Her grade-level team has planning time tomorrow, but she wants to get her ideas together before the three of them meet. She flips through the pages of the teacher's guide of the textbook the team uses, with one student's workbook open beside her. She skims the opening lesson and a few more. She notices the use of arrays and number lines. She reads directions like "ask questions to help them connect the array to the problem" and "ask them to share their strategies with partners." She looks through the many problems the children will be asked to complete as part of these lessons in the student book, noticing the variety of fill-in-the-blank problems, among the typical "solve it" math problems, a few word problems, and one question asking children to analyze where a fictitious person went wrong with their multiplication strategy. She sighs. The students definitely need this work with multiplication, but she thinks about her class and wonders how children will connect to these questions. Mathematics instruction has at times felt like pulling teeth this school year, and she wants to ensure that such important mathematical ideas and representations make sense to them and that they feel they can use these ideas in the future. She wants them to feel connected to the mathematics but also ownership of the mathematics, and to have many, many chances to share their own thinking in meaningful ways. She sets about thinking how she will adjust her curriculum to feel more relevant to her students, but she isn't sure where to start exactly.

INTRODUCING THE STRAND OF KNOWLEDGES AND IDENTITIES

Perhaps you have felt like Alicia, at once eager to ensure your students are learning core mathematical concepts but also not sure you can effectively bridge the mathematics with the students in your class. This chapter lays out key ideas in getting started with the multidimensional processes of making mathematics instruction culturally responsive to children, by addressing the knowledges and identities students bring to the classroom and describing strategies for how to both notice the opportunities to leverage connections that exist in your lesson plans and offer ideas of how to make the connections stronger.

The first strand of the Culturally Responsive Mathematics Teaching Tool (CRMT2) is Knowledges and Identities because fundamentally being a responsive and sustaining educator means working with various forms of deep knowledge of children, their families, and their communities and giving opportunities for that knowledge to shine in the classroom. It is important to understand that each child has mathematical ideas (Carpenter et al., 1999/2014), engages in mathematical practices at home and in their communities (Civil, 2007; Gonzalez et al., 2005), and experiences mathematics in ways that may or may not acknowledge their community, cultural, and intellectual wealth as resources for the mathematics classroom (Yosso, 2005). As teachers, we fundamentally shape *what* children learn and *how* children learn. We shape their identities as math learners (Aguirre et al., 2013; Gutierrez, 2010/2013). This theme reflects the profound importance that knowledges and identities play in mathematics education. This chapter is the first of three chapters with similar structures.

You will learn about the various ways knowledge bases and identities impact the way we teach and learn mathematics in multiple spaces inside and outside the classroom.

BIG IDEAS IN THIS CHAPTER

- Every child engages in mathematics at home, in their communities, and at school. CRMT supports teachers to make connections between home and school practices, by centering community and experiential knowledge.

- Every child's mathematical ways of being should be honored in our classrooms, in ways that humanize children, their families, and communities.

- Every child has mathematical ideas that should be shared, acknowledged, and built upon for mathematical learning.

We will introduce the three dimensions that make up the strand Knowledges and Identities:

1. *Centering Cultural and Community Funds of Knowledge*

2. *(Re)Humanizing Mathematics*

3. *Honoring Student Thinking and Ideas*

Along the way you will learn how to ground mathematics instruction in ways that honor community funds of knowledge of children and their families and attend to humanizing practices alongside honoring students' ideas in the classroom. We will share a rubric for each dimension that includes a description of the dimension, ideas for how to attend to the dimensions in mathematics instruction, and how to interpret the rubric. In addition, every dimension is accompanied by a short teaching story and an activity for your own professional learning.

DIMENSION 1: CENTERING CULTURAL AND COMMUNITY FUNDS OF KNOWLEDGE

ESSENTIAL QUESTION

How does my lesson help students connect mathematics with relevant/authentic issues or situations in their lives?

Our first dimension centers on cultural and community funds of knowledge that children and families contribute to the classroom. This dimension comes first because we want to intentionally place children and their family, cultural, and community practices at the heart of our mathematics instruction. Built on a large research base in education (Civil, 2007; Gonzalez et al., 2005), *Centering Cultural and Community Funds of Knowledge (CFoK)* consists of the everyday knowledge and practices existing in homes and communities that often go unrecognized as intellectual resources. These practices or activities often involve mathematics. Everyone engages in mathematical thinking, reasoning, and actions that traverse many aspects of our lives. Some common examples include creating a family, work, or team schedule; cooking and game-playing with family and friends; organizing fundraisers for community centers, places of worship, and school parent-teacher associations; data analytics of sports; or optimizing transportation routes, budgets, and do-it-yourself projects. Other examples include rationing clean water for daily family needs, shipping and receiving natural disaster (hurricane, fire, flooding) supplies for neighborhood distribution, or making decisions about medical care.

CFoK can also include school-based mathematics learned in different global communities that grandparents, parents, and children may bring with them as a resource. Contrary to popular belief, math is not universal (Perkins & Flores, 2002). Symbols, including decimal points, colons, and commas, can alter the quantity, value, or operation being reflected in the symbolic notation or procedure. It is not just that there is an "American" way and an "everybody else's" way. Rather, algorithms taught in schools differ between countries and systems of schooling and are informed by particular histories in those settings. Sometimes, algorithms used by other countries can be more similar with each other than algorithms privileged in the United States. For example, Figure 3.1 depicts the division algorithm of 127 divided by 4 as shared by one of Julia's students who grew up in Turkey and as performed by Maria's mother (Susana), who is from Peru. Before you proceed, pause and look for the similarities and differences between the two algorithms, and compare them to how you would solve the problem.

Furthermore, numbers can be written in multiple ways. For example, in many places around the world the numeral 7 can look like 7. While this may seem trivial, if you penalize students for using an unfamiliar symbolic notation rather than the "American" way, you send a clear message that diverse ways of doing and representing mathematics are not tolerated in your classroom. Knowing that mathematics can look different around the world can be a resource to broaden our understanding of mathematics as a humanizing activity that occurs in our global community and has similarities and differences to our own understandings.

FIGURE 3.1 Two Algorithms for Division

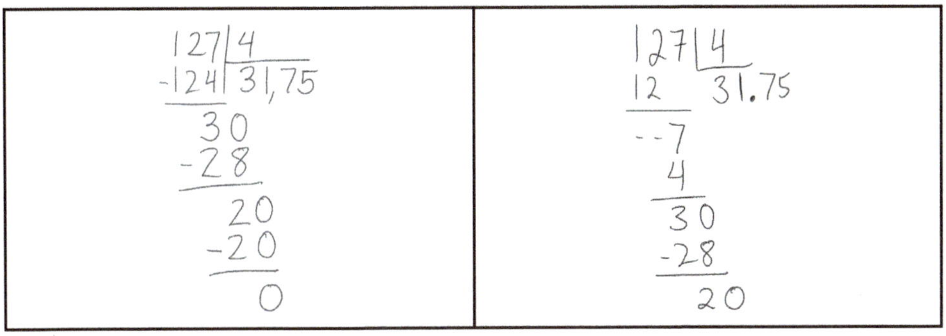

The left is the division algorithm of 127 divided by 4 as shared by one of Julia's students who grew up in Turkey and on the right as performed by Maria's mother (Susana), who is from Peru.

Instructional Approaches to Make Authentic, Not Essentialized, Cultural and Community Connections

Ask any elementary school teacher and they are likely to tell you that their mathematics curriculum is not relevant to their students. But ask them how they make it relevant, and they may offer a range of suggestions. There is no one right way to ensure the curriculum is relevant to your students. However, there are some more productive ways and some less productive ways. The more productive ways support children in drawing on and deepening their understanding of a phenomenon, activity, or tradition that is a part of their lives. They allow children to be the experts on the situation and use that knowledge to help reason through the mathematics of the solution strategy. The less productive ways can at best be surface-level connections to students' lives and at worst can reinforce cultural stereotypes or inflict trauma and violence on children.

Turner et al. (2012) describe how as teachers learn to attend to students' multiple math knowledge bases, they go from making emergent connections, or initial attempts, to making meaningful connections. Emergent connections may be more superficially based on community settings that are presumed to be meaningful for students based on their presence in the community: a pizza parlor, movie theaters, an ice cream shop, the daily visit of the elotero outside of school, or a trip to the bowling alley. But as teachers get more experience with how to draw on the community funds of knowledge, the mathematical tasks themselves shift toward the math that is embedded in meaningful contexts. Some of the most interesting mathematics that emerges from the community comes from engaging with people in the community. This level of engagement helps build connections to the emergent mathematics people face and gives a level of authenticity to community-based math tasks. What is the meaningful mathematics in the pizza parlor, buying pizza or learning how the owner plans for the day? Buying chips from the elotero or tracing his path through the streets of your city?

PAUSE AND REFLECT

- How do you think about relevance? When you work to make mathematics relevant to your class, whose experiences usually get prioritized? What path do you usually take—looking for something relevant to most children or looking to prioritize certain children's experiences?

Any particular activity context doesn't have to be relevant to everyone in the class; rather, it can have deep anchors into the experiences of one or some students, while drawing other students into learning more about it. This is the "mirrors and a lens" concept commonly known in multicultural education: Curriculum can be a mirror, or reflect the experiences of students, or can be a lens, providing them a glimpse into someone else's experience. But more often than not, mathematics curricula reflect a dominant white, upper middle class, heteronormative, gender binary, English-only, seemingly generic reality—a reality that only privileged students may have access to understanding. Therefore, more often than not, you will need to critically review your curriculum and adjust mathematics activities to be more relevant to students and tap into their funds of knowledge.

Keeping in mind the principles outlined in Chapter 1, there is a big picture here that guides our heart-compass when we are not sure whose reality to privilege for one particular lesson or activity. When we take the principles to heart, we can usually find our answers. We can ask ourselves questions such as, *Who is in need of the connection right now? Who has yet to shine?* Certainly, not every lesson is going to feel relevant to all children. One approach we can take is considering the whole school year as our unit of time: Over the school year, commit to mathematizing a variety of experiences from the various communities and families that make up the classroom. Do the heart work necessary to learn about the experiences of a local community for one lesson or unit, then another, then another. This is a way of developing your own critical consciousness (Landson-Billings, 1995). While a particular lesson or activity has its intended connection to a child or community, there is still flexibility and movement, allowing for other entry points and connections.

A variety of strategies exist for doing the consciousness-raising heart work to teach through **CFoK**. Here are three that are absolutely necessary:

1) **Get to know students and their families first:** through casual conversations, genuine interest into students' lives, and home visits that are designed to learn about the family and their many cultural and linguistic resources. You can learn quite a bit to bridge mathematical activities with their lives.

2) **Get to know the community in which you teach:** Learn about the local histories and tribal community connections, as well as local issues and how they impact various stakeholders in the community. Physically explore the neighborhoods near the school; join community organizations including religious or secular organizations; and look for ways you might draw the mathematics out of local issues and action projects.

3) **Practice mathematizing:** when you're out walking, rolling, getting together with friends and family, and taking care of children. Notice and wonder about the world around you and the mathematics in it. Shapes, numbers, angles, patterns, areas, perimeters—the mathematical concepts that emerge from a variety of natural, urban, rural, and suburban settings—are all viable connections to the funds of knowledge of students. And the sharper your own mathematizing lens is, the more you will notice how to connect children's experiences with mathematics and be able to recognize when children are mathematizing their surroundings and experiences.

Understanding the Rubric for Dimension 1

The rubric is intended to be a guidepost for analysis, to help you understand if the lesson has multiple and deep connections to the lived experiences of students from marginalized communities who are present in the classroom or if it is an area that needs modification and improvement (e.g., because there are no connections.) The essential question for *Centering Cultural and Community Funds of Knowledge* is, *How does my lesson help students connect mathematics with relevant/authentic issues or situations in their lives?* Figure 3.2 shows the rubric for Dimension 1.

A rubric rating of 1 suggests that **CFoK** is ignored or addressed only in terms of generic kids' interests. Statements such as "all kids like _____ (pizza, candy, cookies, etc.)" suggest a benevolent intent but harmful impact by dismissing what children and their families might actually know about those contexts. Often, the tasks in textbooks are at a Level 1 as they have been written for a general children's population. As the contexts for mathematics tasks, and the opportunities to feel connected to the central mathematics tasks are made, children find their connections with the mathematics.

For example, children may have different experiences with making and sharing cookies with family members—from the recipes used to the baking tools

Fragile/Margin → Strong/Centered

1	2	3	4	5
There is no evidence of connecting to students' cultural funds of knowledge (parental/community knowledge, student interest). This could include claims of "cultural neutrality" and appeals to a universal nature of mathematics.	There is at least one instance of connecting a math lesson to community/cultural knowledge and experience, such as during the lesson launch. Lesson briefly draws on student knowledge and experience, but they are not central to the lesson. The focus is with one student or a small group of students.	There is at least one sustained episode of sharing and developing collective understanding about mathematics that involves connecting to community/cultural knowledge to analyze authentic situations or issues in students' lives.	There are many sustained episodes of sharing and developing collective understandings about mathematics that involve connecting to cultural/community knowledge (e.g., student experiences are mathematized; student/parent connections with math work; math examples are embedded in local community/cultural contexts and activities such as games, etc.).	The creation and maintenance of collective understandings about mathematics that involve intricate connections to community/cultural knowledge permeate the entire lesson. This would include hook/intro, main activities, assessment, closure, and homework. Students are asked to analyze the mathematics within the community context and how the mathematics helps them understand that context.

available. Thus, it is important to find out through questioning (prior to or during the lesson) what students might know about, for example, sharing, counting, grouping, or dividing treats that might help your lesson design and implementation. The more you know, the more space you create for mathematical connections and understanding with **CFoK**.

Rubric ratings of 2 and 3 reflect attempts to connect to **CFoK**, but the connection is brief or infrequent. This might mean that a lesson has a connection at the beginning, such as a storybook to set the context or a few minutes for the teacher to ask a question about students' experience with a specific activity, only to never bring it up again in the lesson. Ratings of 4 and 5 reflect intentional action of making students' funds of knowledge central to the lesson activities. It can be seen throughout different phases of the lesson and involves students making connections to their own experiences and those of others. A book might be used to introduce a topic, and children may be asked to share their experiences with the context; they are invited to draw on their understanding of the experience as they make sense of the central mathematics tasks; they may be invited to justify their reasoning not only from a mathematical perspective but also in ways that reflect their understanding of the situation. Reflection is key to a rating of 5. This rating requires opportunities for students to explicitly examine the mathematics within a specific community context and reflect how the mathematics they are learning helps them understand and empathize with what is happening in that context. Lessons that tend to be on the higher side of the rubric are often situated in local, specific community sites familiar to children, where the mathematics is actual mathematics one may need to do in the setting and in which children engage in decision-making and justification.

Teaching Story: Centering Cultural and Community Funds of Knowledge

The children came into Ms. D's fourth-grade classroom and were greeted by a picture of a neighborhood hit by Hurricane Maria, the devastating storm that killed over 3,000 people and took out power to the island of Puerto Rico for months. Without sharing the location, the teacher asked, "What do you notice about this picture? What do you wonder?" Children pointed out houses were destroyed, roofs caved in, trees were blown down, there were broken windows and debris everywhere, and there were no people. Some of their wonderings included the following: Was this damage done by a tornado or hurricane? How many people died or were hurt? Where was this? How many homes were destroyed? Are families still homeless? One child recognized the photo and said, "That's Puerto Rico, where my cousin lives." The other children asked if their classmate's family was okay.

Ms. D points out that natural disasters hit our communities every year.

"Turn to your partner. What are some examples of natural disasters that you know about?"

Students began whispering to each other. Words like *hurricanes*, *tornados*, *fires*, and *floods* come out. The teacher asked a few students to share their thoughts. Some additional examples come up:

> "That heatwave happened in summer, lots of people got heat stroke. We did not have air conditioning. Hella hot."

> "Remember the windstorm in November? Our power was out for three days."

> "The fires were really bad this year. We had to pack a bag with supplies in case they got too close."

> "What about earthquakes? Those are scary."

The teacher wrote these examples on the board. "You have a lot of great examples here. I am curious about what Ellis said about packing a bag with supplies to have ready if fires come to close. Could you tell us more?" Ellis proceeds to explain his family's story.

The teacher then asks a more general question: "What can we do to be prepared in case of a disaster? What are your thoughts?"

Students offer ideas: "Make sure we have food and water," "flash light," "move to a place that doesn't have these disasters," "charger," "medicine or bandages if someone is hurt."

Then Jasmine interjects, "But this all depends, like how much food and water? For how long? And for how many people?"

Ms. D is excited; the students have made several real-world math connections already. She states, "You are amazing! You've already started to identify important factors related to our community math project. Today we are going to think about how we can best protect our families and communities if a natural disaster happens. We are going to create a basic emergency kit to help us be safe. We will use our collective experiences and diverse math knowledge to figure out what supplies are needed for the kit. We will make a plan for the emergency kit. We will vote on the 'best plan' for the emergency kit. Then we will write a letter to the principal to buy the supplies and create the kits for every family in the school. As Jasmine has already pointed out, we are going to have to make some important decisions and assumptions for this project. Let's get started with talking about what we mean when we say 'safe.' What does feeling safe mean to you? *¿Que significa sentirse seguro?* Share some ideas with a partner, and I'm going to get our chart paper to take notes."

Though this is just a quick peek into Ms. D's lesson intro, we see some evidence of how this lesson is going to shape up on the **CFoK** rubric.

- Pause for a moment, and turn back to Figure 3.2 (Rubric for Centering on *CFoK*). From the brief glimpse into this classroom activity, what rating (1, 2, 3, 4, or 5) do you see evidence for? Why?

Did you note "at least one episode of sharing and developing collective understanding about mathematics that involves connecting to community/cultural knowledge to analyze authentic situations or issues in students' lives"? The hurricane was impacting at least one child's family, and the invitation to discuss what they understood about natural disasters brought more children and their knowledge into the conversation that they were invited to share ("share with a partner" followed by class discussion). Children were also invited to tap into what makes them feel safe and encouraged to draw on their own experiences as they got started on the task. This is further indication that this lesson contains many sustained episodes of sharing and developing collective understandings that will carry beyond the initial hook or launch of the lesson. This lesson may be on its way to a 5, depending on how it continues to unfold. You might pause and ask yourself, *What decisions might the teacher make from here to strongly center cultural and community funds of knowledge?*

ACTIVITY: Mathematizing Your World

While we may all believe that "mathematics is everywhere," developing the skill to notice and describe the mathematics of particular activities and situations takes practice—practice to name the specifics of what is mathematical in a given situation or what has the potential to be mathematized. In the vignette above, for example, Ms. D heard the potential for mathematical exploration in students' contributions, as they named necessary items and raised questions about quantities. We can imagine that an experienced teacher like Ms. D, who has learned to listen with her heart and not just her mind, was excited at the numerous ways she was hearing and validating students' mathematical and experiential thinking.

We argue that the more practice you have in the specificity of naming mathematics in the world, the more you will notice the connections in your students' ideas and offerings. And that's the kind of mathematical noticing that is at the core of CRMT. So, to sharpen that mathematical lens, choose one or all of the following to try out today:

1) **Get outside.** Take a walk, roll, or slow drive in your neighborhood. Using your senses, what do you notice? What do you wonder? What wonderings could be investigated with mathematics? This is fun to do with a companion and share your thoughts and ideas. The process of communicating your thoughts as clearly as possible helps to sharpen your mathematical lens.

This activity is also adaptable to GoogleEarth or other mapping software that lets you digitally explore an area, but, you don't get the same advantage of being out in the community.

2) **Look at photos.** Flip through a photo repository like a family album or a friend's Instagram feed. Pause and engage in noticing and wondering with some photos. Or, as you reminisce through something like a family album, notice how many of your observations contain mathematical thoughts: "Wow, this must be from at least twenty years ago. Look how much you have grown!" These two comments have quantified elements (at least twenty years ago) and relational thinking (wow, look how much you have grown, from now to then). There's more we could unpack from just these two statements (is the comment about growing an additive or multiplicative comparison?), but ultimately the goal is noticing how your reflections contain valid mathematical ideas.

3) **Reflect on your mathematical day.** Pause at the end of the day and notice the mathematics: You might start by just reminding yourself what you did today. Then, you can look at each item through a mathematical lens. It can be much harder to ask yourself, "Where is math in my life" than to first notice what you spend your time doing, and then unpack the mathematics in there. For example, I might say, "The first thing I did today was take my prescription medication, like I do everyday." Then unpacking the activity, I can notice the mathematics: To take the right dose of levothyroxine, I have to know what I took yesterday, so I have to pay attention to patterns and the passing of time. Then I have to get the right number of pills to make up the required dose, so I'm counting and using precision, and double checking that a differently dosed pill didn't sneak past the pharmacist. I think you get the picture.

4) **Ask families about how they do math.** Make a point to talk with a parent or other family member of a student and ask questions, or share a survey, about how they see mathematics operating in their lives. Reflect on the answers families give and start planning how to use this information for the classroom.

DIMENSION 2: (RE)HUMANIZING MATHEMATICS

ESSENTIAL QUESTION

How does my lesson support creativity, broaden what counts as mathematical knowledge, and affirm positive mathematical identities for all students?

As mentioned in Chapter 1, rehumanizing mathematics broadens what counts as mathematics and how we experience it. It is a reclaiming of mathematics in relation to our own and others' wholeness (Gutiérrez, 2018). In this way, it's complementary to but different from Dimension 1 in that it focuses on the ways mathematics is framed as a human activity, with historical roots in a variety of communities, and adds another way to feel connected to it. If Dimension 1 is about how students' knowledge and communities can be connected to mathematics, this dimension is about how the mathematics itself is presented to children: as a creative versus closed process; as rooted in history within a variety of communities; and as linked to our bodies, not just our minds.

Dehumanized mathematics is a series of disconnected facts or predetermined procedures and processes sanctioned by a dominant group, which is how mathematics is usually presented. Therefore, *re*humanizing mathematics is a creative and dynamic endeavor, inviting diverse thought, innovation, and shared understanding of knowledge. It is rooted in the notion that mathematics, as a knowledge base and set of interconnected practices, must grow and be nurtured by people past, present, and future. Mathematics has social, cultural, and historical components that children need to learn. For example, mathematics is a real-world phenomenon, found in all parts of the globe, and part of everyone's histories, from the abacus and its methods of use in China, to the sexigesimal systems of the Mayans, to threads of the quipu that traversed the Andes as part of the Incan Empire's record keeping, to the geometries, algebras, and numerical calculations used in everyday life across the entire continent of Africa, and more too numerous to list. It also must be acknowledged that school mathematics tends to focus on the contributions of white European cisgendered males. If you do a simple Google image search for "mathematicians," you will see mostly white male faces. You must specify women or Black or Latino/Hispanic or Asian to get other images. Who we see as mathematicians matters. Rarely do mathematics texts acknowledge contributions from civilizations such as these, nor acknowledge that the methods taught in school have roots in particular histories of privilege. We usually do not acknowledge it.

Furthermore, *where* we experience mathematics also matters. Mathematics is place-based, meaning that we can engage with mathematical ideas to better understand how space is historically and culturally constructed, lived in, and reconfigured, affecting access to basic things like clean water, healthy food, civil rights, public transportation, financial services, and green space (Rubel et al., 2016; Rubel & Nicol, 2020). For example, Chao and Jones (2016) recount how young Black children in Ms. Jones's preschool class mathematized a role-play situation depicting Rosa Parks and the Montgomery bus boycott—an iconic event of our country's civil rights struggle for equality. The room was set up as the inside of a bus with four rows of six seats—three seats on each side with a large gap between the first two rows and the last two rows. The 20 preschoolers, all four- and five-year-olds, role-played passengers getting on the bus as the bus driver (Ms. Jones) told them where to go. Some children tried to sit near the front and were told they could not and had to move to the back of the bus. Children were asked to reflect on how this situation made them feel. They voiced feelings of being scared and mad. They felt the situation was not fair. They were able to use early numeracy concepts like cardinality, one-to-one correspondence, and in some cases, counting on strategies to determine that there were enough empty seats available (12) for the eight passengers told to stand in the back. The children voiced that fairness could be achieved if all were able to sit, especially, as one child noted, since each person paid the same amount to get on the bus.

This is an example of how a teacher's lesson can help young children feel empowered to use mathematics to identify injustices and unfairness while affirming historical and mathematical significance in these children's lives. Ms. Jones, a Black woman, was an experienced teacher who knew the importance of teaching what she called "thrival skills," where children learn to simultaneously experience joy and thrive while also learning skills to survive in a racist society. Importantly, her own positionality and her developed sense of teaching in response to the realities of her students contributed to the design of the bus lesson. In turn, the mathematical, historical, and cultural space is rehumanized in this setting for children to grow, learn, and thrive.

We can look to Indigenous scholars to understand another perspective on humanizing mathematics and place. There are some Indigenous communities in which the notion of place-based learning suggests complete interconnectedness—there is no separation between place and people. Storytelling and walking (observation) are forms of "knowing, being, and making" that many Indigenous pedagogies cultivate (Barajas-López & Bang, 2018; Furuto, 2014). For example, Filiberto Barajas-López and Megan Bang (2018) created an I-STEAM summer camp that centered Indigenous ways of knowing and making for elementary and secondary students representing over twenty tribes across the Americas. Young people engaged in a variety of activities including storytelling, clay sculpting, and nature walks. The interconnection among mathematics, nature, people, and culture were apparent as young people created clay pots and basket designs, moving from two dimensions to three dimensions while integrating the patterns observed on nature walks and through conversations with elders and artists into their designs—highlighting "plant-relative" symmetry and repeated groupings. Plants (nature) are considered relatives to people, which impacts how Indigenous youth see, think, communicate, and create in the world. Thus, mathematics is experienced within this wholeness/interconnectedness. For communities who experience a disconnection from place and space, reclaiming our places and spaces in which we live, work, and play is an act of rehumanizing mathematics.

Rehumanizing mathematics also explicitly challenges fixed notions of who can and cannot be "mathematical." It works in the service of honoring children, their families, their histories, and their traditions in the generation and advancement of mathematical knowledge inside and outside of school (Gutiérrez, 2018). This dimension is a direct response to mathematics being a tool used to dehumanize people. School mathematics has historically played a role in ranking and sorting children through the use of standardized test scores and curricular tracking (Berry et al., 2014; Boaler, 2002; Flores, 2007; Gutiérrez, 2008). Our hyperfocus on "achievement gaps" has resulted in children being objectified with labels such as high/low, slow/fast, or disheartening terms like "uneducable" or "bubble kids" (Horn, 2007; Gutiérrez, 2008; Gutiérrez, 2010/2013). Boaler and Selling (2017) have documented the long-term effects of tracking and labeling that left young adults feeling "psychologically imprisoned" (p. 96). These dehumanizing experiences can lead to surveillance, silencing, and isolation of children through acts such as snatching a child's pencil out of her hand, isolating a child from his peers in the back of the room, demanding a child communicate mathematics in a "standardized" way (e.g., English only), or ridiculing a child for asking a question or making a mistake. These acts are an assault on a child's sense of worth and their developing mathematical identities, collectively known as acts of dehumanization.

There are many ways that curricular materials can contribute to dehumanization, chief among them how people are represented in story problems. Fifth-grade teacher Ms. Ross had her students analyze the ways gender was represented in word problems from their curriculum. She asked them to consider, "What does the problem say? What does it mean? Why does it matter: to me, to mathematics, to the world?" (Yeh & Otis, 2019, p. 90).

Within the three word problems the fifth graders examined, they noticed "Word problems with girls' names provide context related to looking pretty, being helpful, and being a homemaker … . Word problems with boys' names focus on sports and competition" (p. 92).

As their analysis went deeper, they began to generalize: There was only one way to be a boy presented here, and another different way to be a girl, and there was no way to be anything else. Further, nonnuclear families were not represented, and only binary opposite-sex relationships were included (i.e., a dance that children are going to). Children were encouraged to question, "Who does this privilege? Who is silenced?" (p. 92). All of this analysis took place alongside mathematical discussions of how fractions and relationships between quantities were also used in the questions.

On the other hand, rehumanizing mathematics can lead to liberation (Freire, 1970/1993), intellectual freedom (Boaler & Selling, 2017), and centering of mathematical agency that embraces connection, creativity, and wholeness. There is an entire collection of examples provided in the NCTM 2018 Annual Perspectives in Mathematics Education titled *Rehumanizing Mathematics for Black, Indigenous, and Latinx Students* (Goffney et al., 2018). In a particularly poignant chapter, Melissa Adams Corral described themes she found in her fourth-grade Latinx students' "mathematics stories"—written narratives about their experiences with mathematics (Adams, 2018). These themes included the power of collaborative learning, use of physical representations such as manipulatives, being able to express their mathematical ideas in multiple ways and in multiple languages, and the positive role of family support in mathematics learning. In their math stories, children articulated powerful images of what helped and hurt their mathematics identities, including their own agency and perseverance of *"no te rindes"* (don't give up). One child, Estrella, shared that in second grade, mathematics was so hard she almost "stayed there" (i.e., retained). However, her memories of fourth grade illustrated a significant shift toward feeling empowered. She loved "the problem of the day," a routine high cognitive demand activity in the class. She felt math was easier and that she had support because her best friend was "by her side." She identified college aspirations as well when she wrote, "I have gotten all As all year. And so far I think I will get into college free" (p. 126).

Estrella's story represents many aspects of her developing mathematics identity. It is important to note that as her math identity changed, complex problems that were once a barrier became a welcome challenge. She had peer support, and she saw the link of success in mathematics to going to college and obtaining scholarships. You see aspects of her past, present, and future—her whole-self—depicted in her math story. Teachers can learn so much from their students through storytelling, an abundant source for rehumanizing mathematics (Barajas-López & Bang, 2018). When children see themselves welcomed and reflected in the regular happenings of their mathematics classrooms, when they are linguistically, culturally, and compassionately embraced, then they are fully humanized in the classroom. Of course, like all dimensions of CRMT, *(Re)Humanizing Mathematics* does not stand alone. You can read more about Melissa Adams Corral's humanizing multilingual mathematics classroom in Chapter 8.

ACTIVITY: Examining Humanity

Look at the following three problems modeled after those commonly found in a popular elementary curricular program.

- Theresa used $\frac{5}{9}$ yard of ribbon to decorate her backpack. Cynthia used $\frac{7}{12}$ yards of ribbon in her dress. Which girl used more ribbon? How much more did she use?

- Ms. Smith noticed that her third-grade classroom has 10 boys and 20 girls. How many times as many girls as boys are there?

- Mrs. Israel is crocheting a scarf for her nephew. She isn't finished yet, and the scarf is $\frac{5}{6}$ yard long and $\frac{1}{3}$ yard wide. What is the area of the scarf so far?

In what way might students find the word problems above relatable? In what ways might they find them dehumanizing? Think about a specific group of children or child you might know in a grade level that might use these problems. How might you adapt the problems to be more humanizing, and more relatable, but work on similar mathematical concepts? To learn more about how Ms. Ross engaged her students in the Say-Mean-Matter protocol, see Cathery Yeh and Brande Otis's article (2019) "Mathematics for Whom: Reframing and Humanizing Mathematics."

Instructional Approaches to (Re)Humanize Mathematics

There are many ways to *(Re)Humanize Mathematics*. We just have to be diligent as teachers in ensuring we are approaching humanization from not just a curricular perspective but also in our instruction. Here are some key ways to build humanizing practices into your instruction:

1. **Recognize that children are whole, perfect, human beings.** First and foremost, keeping the fact that children are whole human beings who need to be loved and honored for who they are in your classroom— this is step one! And yet this is easier said than done. John Henner, researcher and deaf activist, once stated, "My work starts with the radical idea that deaf children are people too." So little attention has been paid to the mathematical learning of deaf and hard of hearing (DHH) children, leading to their dehumanization. A key part of his team's research suggests that the more access DHH children have to language, in particular the language of American Sign Language (ASL), the stronger they are in mathematics. In this regard, the more humanized they are through access to and embracing of their language, rather than being labeled as disabled and pushed to adjust to a hearing person's world, the more gains they show in math (Henner et al., 2021). Being able to pause and ask yourself, "How have I shown my students that I value their humanity in math class today?" is one way we can keep ourselves grounded.

2. **Co-construct classroom norms.** Humanizing mathematics does not take place in a void. Teachers have to consider how the culture of doing mathematics in a classroom sits within the broader culture of the classroom. Classrooms that center on the humanity of children are going to be better prepared to center on that same humanity when it's time for math instruction. Co-constructed classroom norms recognize a diversity of ways of being in the classroom. This supports students in collaborating on mathematics in ways that honor knowledge as shared and resists the idea that individual knowledge is superior or that mathematics should be competitive. Furthermore, a rigid, punitive environment is unlikely to feel like a humanizing place. A flexible, nurturing environment, where children understand not just the expectations but also how the community benefits from them, is more likely to be a humanized space. This can be achieved through discussion and community agreement and a revisiting of agreements when they start to fail the community.

3. **Listen deeply.** Teachers can also listen for the multiple functions in students' contributions. Honoring and validating students' contributions, even reframing for the class how an idea is useful amidst protests that it is not, can help make a classroom a humanized space. Not only teachers but also other children can and should be supported to listen for and frame their classmates' contributions as innovative and unique, as a way to affirm mathematics identities. This requires teachers to go beyond the IRE patterns in speech (Initiation, Response,

Evaluation, then move on) and instead engage with children's mathematical ideas. This also requires teachers to consider how children may be using their lived experiences to make sense of the mathematics in their classroom and looking for ways to validate such knowledge even if it is in need of refinement. For example, the student who shares one sandwich among four people in an equal sharing problem and says that each gets "one half" of a sandwich may still be learning fundamental fraction concepts about naming parts of a whole but might also be drawing on experiential knowledge that when he shares half his sandwich at home with his brother he tears off a chunk and just hands it to him—concluding that anything less than a whole sandwich is half the sandwich. This method of humanizing, of listening for and probing for the logic in the reasoning that comes from experience, even if that logic will be refined with the rules of mathematics, is an important part of validating students' contributions to reasoning and humanizing their mathematical contributions.

4. **Challenge binary and stereotypical thinking.** In terms of curriculum development and adaptation, teachers can also challenge dangerous binary thinking and stereotyping that seek to sort and rank people into neat little boxes. In the example of Ms. Ross's class, the issue there was both the representation of stereotypical gendered activities (e.g., only girls can sew) and the use of binary genders as a tool to sort (e.g., boys and girls). Ms. Ross's use of the Say-Mean-Matter protocol is one way we can support students to engage in analyses of the world around them, notice when binary or stereotypical thinking is happening, and challenge and change the narrative. Supporting children to rewrite mathematics problems or their own stories in ways that account for complexities of humanity is one way we actively challenge binary thinking.

5. **Reflect students in your class.** A final aspect of (re)humanizing mathematics is to analyze the curriculum from which you are teaching. One kind of analysis is the way Ms. Ross focused on how gender stereotypes show up in word problems, and how there are additional aspects of our identities to explore. A key question to ask is, How is the curriculum a mirror, a lens, or a sliding glass door? In what ways are students seeing their own lived experiences, cultural values, and communities reflected back at them? If the curriculum does not depict a diversity of people engaging in mathematics, teachers may need to supplement. A starting point can be to use the biographies of mathematicians found at such sites as www.mathematicallygiftedandblack.com, https://indigenousmathematicians.org/, and https://www.lathisms.org/ (representing Latinx peoples in the mathematical sciences). With these resources, teachers can introduce their class to mathematicians who are not your

stereotypical white, old, cisgendered males but rather from a variety of backgrounds, countries, and cultures.

6. **Tell the history, present, and future of mathematics.** For mathematics to not be seen as a disconnected set of rules, it's important that we tell stories of mathematics that acknowledge the humans who have contributed to its rich history. On the one hand, this means acknowledging that some stories we tell are false; they are a whitewashed version of history, such as saying the Greeks invented geometry rather than saying it's the Greeks' version of geometry that has been passed down by the victors of wars and not, for example, other geometries of other cultures. It also means telling the new stories—about the history of mathematics, its present, and its future. One resource is the book *The Crest of the Peacock: Non-European Roots of Mathematics* by George Gheverghese Joseph (1991/2011). By humanizing mathematicians, through representation of mathematicians from all over the world throughout history, we also adjust our view of mathematics as timeless and unchanging.

More ideas for *(Re)Humanizing Mathematics* can be found in the *Annual Perspectives in Mathematics Education 2018* volume (Goffney et al., 2018). In addition, the book *Reimagining the Mathematics Classroom (2017)* by Cathery Yeh, Mark W. Ellis, and Carolee Koehn Hurtado, with multiple contributions from classroom teachers, complements the ideas presented on teaching mathematics in elementary schools in ways that humanize children, while also being rigorous and content-rich.

Understanding the Rubric for Dimension 2

The rubric of this dimension centers humanizing mathematics learning in the classroom. As the rubric scale increases, children experience mathematics learning as shared, creative, and relational. Children see themselves in the mathematics, and their worth as doers of mathematics is consistently affirmed. The essential question for *(Re)Humanizing Mathematics* is *How does my lesson support creativity, broaden what counts as mathematical knowledge, and affirm positive mathematical identities for all students?* Figure 3.3 shows the rubric for Dimension 2.

A rubric rating of 1 suggests a classroom climate that is depersonalized with children working in isolation, afraid to make mistakes, and with few if any relevant or relatable contexts in the curriculum. A rating of 2 means that there is some evidence of humanizing practices, maybe evident in how a teacher launches a lesson with a question about what children know about particular mathematics concepts and procedures, or asking a child to share how his or her parents showed them how to solve a problem, or invites

FIGURE 3.3 Rubric for Dimension 2: (Re)Humanizing Mathematics

Fragile/Margin				Strong/Centered
1	2	3	4	5
There is no evidence of humanizing practices. This could include mathematical knowledge treated as impersonal and unquestionable; mathematics of only the dominant school culture in the United States; and lack of connection to students as human beings.	There is some evidence of at least one aspect of humanizing practice in part or all of the lesson, which could include incorporating cultures and histories of students in the classroom, support for physical and emotional components of mathematical knowing, and students taking ownership of ideas or being asked to analyze/question mathematics as presented.	There are some instances of shared and collective construction of knowledge that • expands traditional notions of who can be good at mathematics, • may honor students' histories and cultures, • in other ways affirms mathematics identities across student groups, or • prompts students to be asked to analyze/question mathematics as presented.	There are many instances of shared and collective construction of knowledge that expands and challenges traditional notions of who can be good at mathematics and honors students' and/or marginalized people's histories, cultures, and perspectives *in service of* affirming mathematics identities.	There is a deliberate and continuous presence of humanizing practices, such as students drawing on many different knowledge bases to contribute to the construction of mathematical ideas, honoring of students' histories and different ways of knowing, in particular students from marginalized communities, as well as other forms of affirmation of mathematics identities.

children to critically analyze the contexts of the word problems in their activity book. These interactions are short and possibly limited to the opening or closing discussion in a lesson, or they are invited but not followed up on. A rubric rating of 3 acknowledges an explicit attempt in a lesson to expand understandings and representations of who can do mathematics. This might be done by noticing and showcasing a variety of children's work on the class walls or in the halls or highlighting the hard work of a child that shows a unique way to solve a problem. Or teachers might pull out attempted but unsuccessful mathematics strategies that showcase a productive way of thinking. A rating of 3 might reflect a few moments in the lesson where children learn about how methods were developed by different cultures or are supported to make connections to number systems and symbolic notation used across the world.

A rating of 4 or 5 demonstrates purposeful design across a math lesson that engages the class in developing shared understanding of mathematics as it relates to these (re)humanizing math practices. Students frequently get opportunities to discuss historical, cultural, and community contexts with mathematics. Children's math identities are affirmed throughout the lesson through participation structures that support the exchange of diverse ideas. Student contributions, including their lived experiences and knowledges, are openly valued as resources by the class. Every student is seen in light of their full humanity in all they bring.

Teaching Story: (Re)Humanizing Mathematics

Ethan is an African-American boy in kindergarten and often seeks attention to be challenged intellectually. The class is working on a math modeling problem about packing a lunch box with just enough food so that little is wasted but also so the student in the problem doesn't go hungry. Ethan is selecting how many string cheeses to pack in a lunchbox. He laughs as the teacher approaches and writes "100" in the box. Knowing this student, the teacher knows that this is one way Ethan shows he can play with large numbers. She laughs with him, they talk about how funny it would be to pack 100 cheeses but also how the lunchbox might explode! After they discuss his idea, she asks how many baby carrots, to which he says "60!" After writing 60 in the box, the teacher says he is doing a great job matching his numbers to his writing and says she is looking forward to how he adds all those numbers up.

In this example, the teacher knows to take Ethan's humor as part of his expression. She can focus on his mathematical ideas and make space for him to express his way of thinking about mathematics, by honoring his bid to be seen as a child who can work with large numbers. She explicitly chooses to see his activity as valid and valuable, instead of what another teacher might see as disruptive or off task. In this way, she affirms his identity as a competent doer of mathematics in ways that affirm other parts of who he is, as he engages through play and experimentation.

ACTIVITY: Roots of Humanizing Mathematics

Humanizing may be a new term to some of us. It's important we reflect on ways that we have contributed to dehumanizing mathematics and commit to rehumanizing it.

Knowing how various peoples from across the world created mathematics that was useful and applicable to their lives can give us a stronger sense of just how mathematical people are. To quote Luis Ortiz-Franco, "Chicanos have math in their blood." Put another way, children are the legacy of their ancestors, and everyone's ancestors, including the Mayans of Mexico, the Inca of Peru, and all the various Indigenous civilizations that have cared for Turtle Island since time immemorial, ALL did, and still do, mathematics. As pointed out by Joseph (1991/2011), the continents of Africa, South America, Australia, and Asia are also birthplaces of mathematics. Everywhere that there has been life there has been mathematics! Bridging the past with the present, we can support ourselves and our students to see how we are truly our ancestors' wildest dreams, mathematical hearts and all.

For this activity, take time to learn more about a part of the world that you have personal connections to. What mathematical activities do present-day people, including people close to you, engage in? How are these processes and practices linked to your past? How are these processes and practices influenced over time by other systems: financial, social, educational, and so on? Summarize some of your thoughts as specifically as you can and share them with other teachers engaged in the same activity.

Another activity is to visit the websites www.mathematicallygiftedandblack.com, https://indigenousmathematicians.org/, and https://www.lathisms.org/. What strikes you about the stories of mathematicians featured across these websites? How are your own experiences with mathematics similar and different? Taking time to reflect on the biographies of mathematicians from websites like these can be useful to expand your own understanding of what mathematicians look like, where they are from, and what they are working on.

DIMENSION 3: HONORING STUDENT THINKING AND IDEAS

ESSENTIAL QUESTION

How does my lesson create opportunities to elicit, express, and build on student mathematical thinking in multiple ways (e.g., through gesture, pictures, words)?

This dimension stands on the copious research base that shows all children have mathematical ideas and engage in mathematical thinking from very young ages (Carpenter et al., 1999/2014; Empson et al., 2011; Turner & Celedón-Pattichis, 2011; Webb et al., 2014). Our job as teachers is to elicit student thinking to better understand the ways in which students are making sense of problems, their reasoning strategies, and conceptual and procedural understandings. It furthers the line of reasoning presented in dimensions one and two, that all children bring their brilliance into the mathematics classrooms and that we as teachers must make space to honor and humanize their experiences and recognize when they make connections between those experiences and mathematics.

Students also communicate their ideas in various ways. Being able to identify different ways of expressing mathematical conjectures and other types of viable arguments—either verbally or nonverbally (e.g., through gestures) is critical to a culturally responsive math lesson, especially for multilingual children and emergent readers (Celedón-Pattichis & Ramirez, 2012; Chval et al., 2021; Moschkovich, 1999; 2002). As students learn mathematics in multiple languages, they traverse multiple language registers (mathematical and everyday) in two or more languages. This may result in children expressing their understandings via pictures, pointing, or play. Giving children multiple ways to share their ideas with you and their peers facilitates deeper connections with mathematics and its applications.

Students also develop their mathematics identities through social interaction. They learn quickly whose ideas are valuable and if their ideas are considered important. Cultivating a classroom where students are encouraged to share ideas, understand each other, and empathize with each other creates a space for students to see themselves as valuable and understood mathematical people. In classrooms where students contribute to a positive classroom culture and take charge of the mathematics discussions, all students can see how what they bring to the classroom is uniquely valued not just by the teacher but by other students as well.

Instructional Approaches to Honoring Student Thinking and Ideas

First, mathematics lessons must be structured to provide opportunities for students to access the ideas and share their thinking about them. A classroom in which children are not asked what they think will not be able to meet this dimension on any level. Second, school mathematics tends to hold a place of power over everyday home and experiential knowledge. This difference is made more acute when the cultural ways of being in the classroom are normed to whiteness and students themselves come from diverse communities. Valuing all students' ideas then means that as teachers we must recognize when students' contributions or ideas are different from what is normalized in mathematics but still valid. This requires us to listen carefully to the ways students reason with ideas and numbers. A clear example of this comes from William Tate's (1994) groundbreaking commentary on the centricity of whiteness in mathematics assessment items. Tate wrote of a middle school's mathematics teachers' experiences learning about test answer discrepancies that fell along racial lines:

> A large number of students at their predominantly African American school responded "strangely" to an assessment item on a district wide mathematics test. The basic structure of the test item was as follows: It costs $1.50 each way to ride the bus between home and work. A weekly pass is $16. Which is the better deal, paying the daily fare or buying the weekly pass? When school officials questioned the students about their responses, they found that many of the students had centered themselves in the solution process. These students converted the "neutral" context of the problem into their own property. For instance, the students commented on the fact that more than one family member could use a weekly pass. They also mentioned the option of using the weekly pass on Saturdays and Sundays. In the families of many African American students, the financial providers hold several jobs—both on weekdays and on weekends. For these students, choosing the weekly pass is economically appropriate and mathematically logical. (p. 480)

As he argued, a failure to account for children's cultural understanding led to many students, specifically African American students, getting the math problem wrong, since mathematically the option to buy individual tickets was cheaper when the assumption was that the bus was ridden two times a day, by one person, going to and from one job. But students reasoned that a bus pass that could be shared with others and used for multiple trips a day, was the better value. When the students' thinking was probed, their logic was clear and centered on their own experiences.

Taken in the broader context of the goals and principles of CRMT, we also view this dimension through the lens of children as sensemakers in that their contributions and mathematical ideas should be taken as signs of their current conceptualization of ideas. As teachers, we have been socialized to hear a student's idea and decide if it's right or wrong, or, say to oneself "this child is missing something." We may have been explicitly trained to focus on what children did wrong in a math problem, as a way to help them do better. But a singular focus on what children do not understand detracts from what they actually do understand and are communicating about their understanding. These understandings are important building blocks for making important mathematical connections. We encourage teachers to listen for the ways students' ideas are valuable contributions to the dynamic and collective sensemaking of the class, and not always the contributions we are expecting. Teachers can practice listening closely to students' thoughts, asking probing questions to gain clearer understanding, even if we think we know what is going through their heads. We can also probe for understanding around contributions that we may not see initially as mathematically valuable. Sometimes children say things that seem designed to derail us and our lesson goals. But seen through a humanizing lens, we are encouraged to instead listen closely to the ideas and needs expressed through their contribution and validate them in the process.

Understanding the Rubric for Dimension 3

This dimension is all about collective meaning making. The rubric for this dimension is focused on how teachers make opportunities for students to share their ideas and respond to each other's ideas. Multiple forms of communication are valued. Multiple opportunities for communication are key. The essential question for *Honoring Student Thinking and Ideas* is, *How does my lesson create opportunities to elicit, express, and build on student mathematical thinking in multiple ways (e.g., through gesture, pictures, words)?* Figure 3.4 shows the rubric for Dimension 3.

A rating of a 1 represents teacher-centered instruction in which the teacher does all the talking. Level 2 requires evidence of some sharing and discussion among students but is limited to just particular students. We might see a Level 2 in a classroom where only one small group of students is talking and the teacher draws on this group's ideas to facilitate a whole-class discussion. Often, status issues are present, since it is likely that a few students holding higher academic status may control the conversation. Level 3 is where the shift is made to a whole-class invitation for sharing thinking and every student is involved, with at least two different strategies for eliciting ideas from students, such as pair-sharing or sequenced strategy sharing. Levels 4 and 5 are reached when student thinking is clearly the focus of the lesson. Ideas are elicited across all lesson phases, from initial hook to closing discussion. There is clear evidence that all contributions are valued, which could mean the teacher and students are working to understand how even in a "wrong" answer a student may be making sense of some important mathematical idea. The key difference between 4 and 5 is that Level 5 means that essentially a whole lesson is permeated with sensemaking and refining of ideas between a large variety of students and teacher, as well as students taking the role of eliciting ideas from each other (e.g., through question asking, collaborative learning).

FIGURE 3.4 Rubric for Dimension 3: Honoring Student Thinking and Ideas

Fragile/Margin → Strong/Centered

1	2	3	4	5
The lesson does not include attention to student thinking.	The lesson includes some attention to student thinking.	The lesson includes at least two strategies aimed at making student thinking public.	The lesson includes multiple strategies to make student thinking public.	The lesson includes multiple strategies to make student thinking public.
Mathematical contributions in the lesson are almost exclusively from the teacher.	Teacher elicits student thinking of an individual student or small subset of students.	Teacher elicits student thinking among students in at least one phase of the lesson (launch, explore or summary).	Teacher elicits mathematical thinking across all phases of the lesson.	Teacher and students elicit mathematical thinking across all phases of the lesson.
Shared understanding or collective meaning making is absent.	Sharing of mathematical ideas is among a few select students or between a student and the teacher.	Shared understanding about mathematical ideas and contributions are evident in at least one part of the lesson.	Multiple forms of student mathematical contributions are encouraged and valued by teacher and students.	All contributions are valued and respected by teachers and students.
	Shared understanding is minimal.		Shared understanding between teacher and students as well as among students is evident across the lesson.	There are multiple and sustained opportunities for teachers and students to collectively respond to each other's thinking and contribute to refining mathematical ideas core to student learning.

Teaching Story: Honoring Student Thinking and Ideas

In a third-grade classroom, Ms. Ito circulates as her students solve this problem: "Nora baked twelve cookies with her abuelo. She wants to divide them equally between herself, her brother, and her two cousins. How many cookies do they each get?"

Ms. Ito stops by the desk of Nora, who has a question. "*Maestra*, are these cookies big or small?"

Ms. Ito asks, "Hmm, does the problem tell us if they are big or small?"

"No, but if they are big I can't have too many. And my cousins are little and can't eat a lot of cookies," responds Nora. "I bake a lot of chocolate chip cookies with Abuelito, and they are not too big; usually my mom says we can have two."

Now Ms. Ito knows that Nora has centered herself in the problem and is drawing on her life to make sense of it. Ms. Ito's goal in presenting this problem is for students to grapple with partitive division, so she is expecting that students will notice the phrase "divide them equally" and give each person in the problem three cookies. But she is also interested in the strategies and reasonings they bring to these situations. She has to strategize how to balance Nora's sensemaking with the mathematical goals for the activity. Ms. Ito gets curious about Nora's reasoning: "How would that work out, if you do what you usually do?"

Nora draws two small circles in each of the four circles on her paper. "Then we each get two. That's eight cookies."

"So, eight cookies eaten all together. Is that what you have there?"

"Yes."

"Let's see what the problem is saying. Here, it says we have twelve cookies. Does your model show twelve cookies?"

Nora thinks for a moment, and then draws four more small circles off to the side of the paper.

"Why are those on the side?" asks Ms. Ito.

"Because we are saving them, maybe for tomorrow."

Ms. Ito thinks for a moment. She recognizes that Nora did answer a mathematical question. She answered a quotitive division question of eight cookies and each person gets two, illustrating how four people would get them. Or did she also answer another partitive division question, of eight cookies shared among four people? She decides that there is something to share with other students here and responds to Nora, "I see just how you are thinking about this. So how many cookies did you share?"

"Eight, and four left over," replies Nora, gesturing first to the cookies in the circles, and then to the cookies on the side of her paper.

"So, you showed how to share eight cookies equally among four people, didn't you?"

"I guess so ..." begins Nora, hesitating. "I think I did?" She says doubtfully.

"I think you did too, because you told me how you, your little brother, and the cousins each get two. So how many cookies did you share?"

"Eight," replies Nora, confidently.

"Okay, perfect. Can I share the work you did on this problem with the class, Nora? I want to show everyone how you made a beautiful drawing for sharing eight cookies equally and saving some for later."

"But it's not the right answer," Nora says, confused.

"But it is *a* right answer to a different problem that I think you have a lot of experience with at home; is that right?"

"Maybe. I mean, I have to give out the cookies," replies Nora.

"I see. Now, sometimes math problems ask us things that are just a little different than we would do in our own lives," continues Ms. Ito. With some more prompting, Nora realizes what the question is asking and works her way to the answer. But Ms. Ito doesn't want to leave Nora's great work behind. She asks, "You know, I really appreciate how you used all your knowledge of sharing cookies as you worked through this problem. I bet other people in our classroom had similar ideas. Can we share your first idea with the class, about sharing eight cookies with four people? I think we can challenge the class to tell us what question you were answering."

"Okay," says Nora, and she turns to draw chocolate chips in her cookies.

In this lengthier example, we wanted to illustrate how teachers probe and explore students' thinking, listening, and making sense of the various ways children are centering their experiences as they work to solve problems. We notice that Ms. Ito had a few strategies for eliciting and responding to Nora's ideas, in particular probing around her personal experience with the situation, asking her to show drawings to match her thinking, and inviting her to see how her reasoning applies to a different situation before asking her to apply that thinking to this situation. In a classroom where teachers interact with students in these ways, student thinking and ideas can be honored and respected in the course of mathematics instruction.

ACTIVITY: Learning to Listen for a Diversity of Ideas

We know as teachers that it is easy to notice what students do wrong and what is problematic about their sensemaking. We are trained, in many ways, to notice computational errors and so-called "misconceptions." But as equity-focused educators know, the words we use to describe students matter. We need to unlearn noticing what children do wrong so we can fix it, and we need to relearn to listen to what children understand and connect to it. Further, our sense of correctness is culturally situated. In mathematics, questions about "best buys" usually privilege a perspective where the context of the purchase doesn't matter as much as the cheapest price, but depending on your orientation to what is "best" you may find that you answer a best-buy question in a way that the task author didn't have in mind (for example, choosing to support small business owners over the cheaper price). It can be hard to notice how our own cultural understandings inform our mathematical ideas, but it is important that we do notice the influence of our own experiences on our interpretations of situations, so that we can listen broadly, understand, and validate students' diverse ways of thinking in the mathematics classroom.

Part 1: How Are Cultural Values Reflected in Students' Mathematical Thinking?

Refer to the teaching story above. What is it that the student understands about sharing? In what ways does her understanding reflect cultural values or lived experience? In what ways does it reflect mathematical understandings about equal sharing problems? How might you have approached the situation …

- so that the child felt understood, while also sharpening her ideas of mathematics and what is meant by "equal sharing"?

- so that the class understood their classmate had a good idea, while also learning to distinguish meanings of equal sharing from other kinds of sharing?

Part 2: How Do We as Teachers Learn to Notice and Validate Differences in Cultural Values That Show Up in Mathematics Thinking?

1) Self-reflection: Think of a time when you were misunderstood. How did it make you feel? How did you work through to be understood?

2) Think about what it means for children to be misunderstood, not because they didn't understand the question, but because they focus on values different from those traditionally valued in the mathematics classroom. Are there examples from the texts you teach from? Are there examples from interactions between students in your class that you can remember, where a student was judged for thinking differently? As you reflect on those experiences, consider how you would navigate them as a teacher listening broadly and generously for how cultural values are at play in mathematical solution strategies.

CONCLUSION

In this chapter, we introduced the first three dimensions of the CRMT2. Together they make up the strand of Knowledges and Identities, the first and arguably foundational area of CRMT. We lead with these dimensions because of how central they are to understanding CRMT in its entirety. *Centering Cultural and Community Funds of Knowledge* means we learn deeply about families, we learn to connect mathematics from the world around us to the classroom, and we design and adapt curriculum to center the lived realities of children in our classrooms. *(Re)Humanizing Mathematics* requires us to critically reflect on how we bring the human-centered nature of doing mathematics back into our instruction, which is often dehumanizing for children. And finally, *Honoring Student Thinking and Ideas* means we commit to eliciting student's ideas and seeking to understand and honor their ways of making sense of the mathematics tasks they engage with.

DISCUSSION QUESTIONS

- How have students' knowledges and identities played a role in your mathematics instruction so far?

- What ideas in this chapter resonated/stretched your thinking about mathematics instruction?

- What activity box ideas will you initiate as part of your professional learning journey in culturally responsive mathematics teaching?

CHAPTER 4

· ·

RIGOR AND SUPPORT

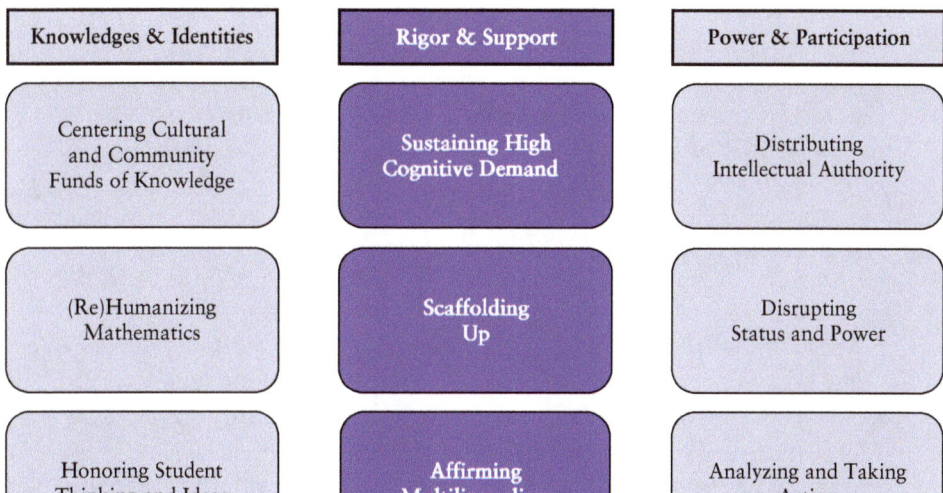

Knowledges & Identities	Rigor & Support	Power & Participation
Centering Cultural and Community Funds of Knowledge	Sustaining High Cognitive Demand	Distributing Intellectual Authority
(Re)Humanizing Mathematics	Scaffolding Up	Disrupting Status and Power
Honoring Student Thinking and Ideas	Affirming Multilingualism	Analyzing and Taking Action

Let's revisit teacher Alicia, who we met at the beginning of Chapter 3:

It's been a few weeks since Alicia worked through her initial dilemma of how to connect mathematics to her students. She has been modifying problems in her curriculum to build on student interests and experiences and initiated a weekly open problem-solving practice. She has also learned to listen more deeply to her students' reasoning, especially the ways they draw on their own experiences to decide what steps to take to solve math problems. Throughout this process, she has made space to talk more with her students about how they are thinking, and she is learning to prioritize their reasoning in addition to looking for accuracy in their solutions. She feels the door is opening more for her mathematics instruction to be humanizing and freeing. And, like with any change, new dilemmas emerge. For example, now that she is prioritizing more open-ended problem-solving, she notices some students shut down after the problem is read or get frustrated if the answer doesn't come fast. She also finds some students are early finishers and need additional challenge. And now that her mathematics instruction is more language-rich, she finds herself doing more translation work for her emergent multilingual students. She discusses these new dilemmas with her grade-level team. In one of their meetings, she says, "I am just having a hard time learning how to get Isaul, my one newcomer kid, into the problem and then leave him to do the work. I can easily spend ten minutes just sitting with him, translating materials into Spanish and English, acting out ideas, but then I lose track of where the rest of the class is. I know I can do better to position my students for success. I'm just wondering how much help is enough, versus how much help is too much?"

It's a big dilemma to know how and when to step in or step out, so that children feel the freedom and support to engage in rich, rigorous, and relevant mathematical tasks, especially when it comes to multilingual students. How can their wealth of knowledge take center stage? Alicia's dilemma is one you may face as well. In this chapter, we take on the strand of Culturally Responsive Mathematics Teaching (CRMT) concerned with centering children's opportunities to engage in rich, rigorous, and relevant mathematics tasks in ways that provide "just right" support that positions them as competent, agentive problem-solvers. This strand is all about how learners are moved from the margin to the center, with strategic scaffolds that maintain high cognitive demand, while viewing their linguistic identities as assets rather than deficits in the mathematics classroom.

INTRODUCING THE STRAND OF RIGOR AND SUPPORT

Strand 2 builds on the foundational ideas in Strand 1. Research has found that there are three important factors that have a positive impact on mathematics success, especially in schools that serve working-class and poor communities: 1. strong family-school/teacher-student relationships, 2. high quality content, and 3. high expectations (Chval et al., 2021; Kitchen et al., 2007). You must have all three. Strand 2: Rigor and Support focuses on access to challenging mathematics (rigor) while maintaining high expectations for all students through content scaffolding and attention to language resources children bring to and develop in the school.

- Children must engage with mathematical tasks that sustain high cognitive demand, while being provided space to grapple with ideas in creative and productive ways.

- Scaffolds can and should provide students with multiple access points and opportunities for deep analysis of math concepts, skills, and strategies.

- Multilingual students must be centered as full participants in the classroom, where their ideas and contributions can be part of the collective mathematical work of the class.

We will introduce the three dimensions that make up the strand Rigor and Support:

4. *Sustaining High Cognitive Demand*

5. *Scaffolding Up*

6. *Affirming Multilingualism*

Along the way we will describe the significance of *Sustaining High Cognitive Demand* while providing intentional supports that scaffold students up to the mathematics, with specific attention to creating spaces that honor and support multilingualism, which is a key way to position the growing population of multilingual students as important mathematical thinkers, doers, and communicators. As in Chapter 3, each dimension is described and accompanied by ideas for how to attend to the dimension in mathematics instruction. We then provide interpretation of the rubric and a short teaching story. We complete each section with an activity for your own professional learning designed to help you develop skills and knowledge in relation to the dimension.

Dimension 4: Sustaining High Cognitive Demand

ESSENTIAL QUESTION

How does my lesson enable all my students to closely explore and analyze math concepts(s), procedure(s), and problem-solving/reasoning strategies?

A key concern when we teach for equity is how to support students in accessing rigorous mathematics. In particular, Black and Brown children and children from working-class communities have the least access to challenging and engaging mathematics. Thus, this dimension is focused on the quality of the mathematical activity students will be engaged in and the commitment to holding high expectations for all students. Access has typically focused on how to ensure that students understand what is being asked of them in a particular task or problem, have opportunities to solve it, and then have methods for communicating their solution(s). But we take this idea further within cognitive demand. When we say access to rigorous mathematics, we mean mathematical activities that require students to engage in a variety of mathematical practices that include but are not limited to understanding and/or articulating a problem, identifying and carrying out a plan, connecting mathematical ideas and procedures, and communicating mathematical thinking to others (Smith & Stein, 1998). This dimension is concerned with ensuring that children have access to high-cognitive-demand tasks, and as they work on the task, they are supported to sustain their engagement and build their math stamina.

We might contrast this perspective with some approaches to differentiation. In some approaches, children are given mathematics activities that are entirely different from the work of their peers in the name of meeting them where they are. For example, some students might get skill-based worksheets, while others get more complex problem-solving tasks. This approach might be based on the idea that math facts must be mastered first before engaging in complex problem-solving. While this view is outdated, it is nonetheless common. Culturally responsive mathematics teachers do not view differentiation in this way. Rather, we first work to provide challenging mathematics to all students, with differentially planned supports based on students' needs. We reserve differentiation for when it is genuinely in response to learning differences and not due to an assumption of lack of ability or missing skills. Choice of activity is certainly encouraged, so long as the integrity of the task is maintained and the cognitive demand is not compromised.

Certainly, cognitive demand is not a new concept. However, we have to keep bringing it up in the context of CRMT in part because children from minoritized communities continue to be denied access to rigorous mathematics—they continue to be socialized into mathematics through low-cognitive demand activities. They don't get to analyze or have the freedom to think.

"High cognitive demand tasks require students to draw on community, situational, and other kinds of knowledge as they work through the task."

We also take up the idea of *Sustaining high Cognitive Demand* as a way to engage the multiple knowledge bases of children. High cognitive demand tasks require students to draw on community, situational, and other kinds of knowledge as they work through the task. Gutstein's (2006) framework of the classical, community, and critical knowledge in mathematics can be a guide for how to expand what counts as *Sustaining High Cognitive Demand* from a CRMT perspective, as we maintain the rigor in the work but also broaden what counts as legitimate knowledge for problem-solving to include cultural and critical ways of knowing. For example, the more children are asked to draw on their community and cultural knowledge to inform how they solve problems, as we saw in examples in Chapter 3 (Strand 1, Dimension 1), when that knowledge is used to solve challenging and novel mathematics problems, the more high cognitive demand is sustained.

Stein et al. (2000) developed a task analysis guide ranging from low-demand tasks, such as memorization and repeating back information, to high cognitive demand tasks that require nonroutine thinking and connecting ideas (see Figure 4.1). Readers of this book might be familiar with diagrams like the one here from Stein et al. (2000), reproduced in Boston and Smith (2009), describing various levels of cognitive demand. Ultimately high cognitive demand is about creating opportunities for students to engage in sensemaking, to connect mathematical ideas and to do the hefty mathematical work themselves. The significance of cognitive demand is in the opportunity for connecting concepts and procedures and developing ideas of mathematics as dynamic and connected (Ni et al., 2018).

Any particular mathematics lesson or activity can hold a variety of levels of cognitive demand. As Boston and Smith (2009) and others have pointed out, cognitive demand is not just about the potential in the task but also about how the task is implemented. While a task may hold potential for high cognitive demand, if the teacher steps in and overscaffolds or does the mathematics for the students, then the cognitive demand is lowered. We discuss this more in the next dimension, *Scaffolding Up.*

Unfortunately, or by design, engaging in mathematics tasks that push for high cognitive demand is not the reality for most students from communities of color. In an effort to "meet them where they are," schools can take a narrow perspective on what it means to be good at mathematics and give children rote memorization and repetitive tasks in the name of learning "the basics." This belief is often associated with social class and race and is enshrined in highly scripted direct instruction texts marketed to schools with high populations of children impacted by poverty and/or Black, Brown, and Indigenous youth. A variety of educational technology that purports to support mathematical learning relies on a similar logic of repetition to solidify learning. Research suggests that repetition makes for learning how to repeat a process; without understanding how that process works, children are less prone to catch their own mistakes and have no way to reason through a solution when a procedure is not yet memorized (see, for example, Behrend, 2001). What we know from research is that children from diverse income, language, racial, and cultural backgrounds can and do engage in complex

FIGURE 4.1 The Task Analysis Guide

LOW-LEVEL COGNITIVE DEMAND	HIGH-LEVEL COGNITIVE DEMAND
Memorization tasks • Involve either producing previously learned facts rules formula or definitions, or committing facts rules formulae or definitions to memory. • Cannot be solved using procedures because a procedure does not exist or because the time frame in which the task is being completed is too short to use a procedure. • Are not ambiguous—such tasks involve exact reproduction of previously seen material and what is to be reproduced is clearly and directly stated. • Have no connection to the concepts or meaning that underlie the facts, rules, or formulae, or definitions being learned or reproduced.	*Procedures with connections tasks* • Focus students' attention on the use of procedures for the purpose of developing deeper levels of understanding of mathematical concepts and ideas. • Suggest pathways to follow (explicitly or implicitly) that are broad general procedures that have close connections to underlying conceptual ideas, as opposed to narrow algorithms that are opaque with respect to underlying concepts. • Usually are represented in multiple ways (e.g., visual diagrams, manipulatives, symbols, problem situations). Making connections among multiple representations helps to develop meaning. • Require some degree of cognitive effort. Although general procedures may be followed, they cannot be followed mindlessly. Students need to engage with the conceptual ideas that underlie the procedures in order to successfully complete the task and develop understanding.
Procedures without connection tasks • Are algorithmic. Use of procedures either specifically called for or its use is evident based on prior instruction, experience, or placement of the task. • Require limited cognitive demand for successful completion. There is little ambiguity about what needs to be done and how to do it. • Have no connection to the concepts or meaning that underlie the procedure being used. • Are focused on producing correct answers rather than developing mathematical understanding. • Require no explanations or explanations that focus solely on describing the procedure that was used.	*Doing mathematics tasks* • Require complex and nonalgorithmic thinking (i.e., there's not a predictable, well-rehearsed approach or pathway explicitly suggested by the task, task instructions, or a worked-out example). • Require students to explore and to understand the nature of mathematical concepts, processes, or relationships. • Demand self-monitoring or self-regulation of one's own cognitive processes. • Require students to access relevant knowledge in working through the task. • Require students to analyze the task and actively examine task constraints that may limit possible solution strategies and solutions. • Require considerable cognitive effort and may involve some level of anxiety for the student due to the unpredictable nature of the solution process required.

SOURCE: Reproduced from Stein et al. (2000) in Boston & Smith (2009).

problem-solving from a young age. Automaticity of facts is part of a developmental process but does not come before complex problem-solving; rather it can be developed through problem-solving. Therefore, we reject this informal mathematical caste system. All children need access to rich, engaging, rigorous tasks when learning mathematics.

> "Automaticity of facts is part of a developmental process but does not come before complex problem-solving; rather it can be developed through problem-solving."

Fundamentally, we have to ask ourselves this question. How can we expect students to be successful in high cognitive demand tasks if we never give them the opportunity to engage *with* those tasks? Further, how can we sustain the high cognitive demand of tasks if those tasks are not relevant to the children working on them? As culturally responsive educators who embrace ideas of designing for learning, we need to focus on providing rigorous and challenging mathematics and then supporting access as described in two related dimensions that follow (*Scaffolding Up* and *Affirming Multilingualism*).

Instructional Approaches to Sustaining High Cognitive Demand

Some of us might have learned to associate long word problems with cognitively demanding tasks, maybe because when we think of what challenges our students, it is a long, wordy, multipart problem. But length of the problem is not the measure of cognitive demand. Determining cognitive demand is about the mathematical thinking and communication that is required in the task. For example, a long word problem may introduce linguistic demand, in particular for multilingual learners, but may not be mathematically challenging. In general, tasks that are open ended—that have both multiple entry as well as multiple exit points—have higher cognitive demand.

One way to regularly include accessible high cognitive demand tasks to engage students is with mathematical modeling set in cultural and community contexts. Mathematical modeling is a cyclic process that helps students analyze real world situations (Garfunkel & Montgomery, 2019). It emphasizes sense-making, decision-making, building mathematical models, interpreting, revising, and sharing their models to help others (Arnold et al., 2021; Suh et al., 2021). When modeling tasks are grounded in children's experiences, activities, and communities, there are more opportunities to leverage their multiple mathematical knowledge bases,

> "When modeling tasks are grounded in children's experiences, activities, and communities, there are more opportunities to leverage their multiple mathematical knowledge bases, tapping into their understanding about the context and mathematical relationships to solve a novel problem (Anhalt, 2014; Turner et al., 2021)."

tapping into their understanding about the context and mathematical relationships to solve a novel problem (Anhalt, 2014; Turner et al., 2021). For example, students might investigate how many plastic bags are needed to braid jump ropes for the PE class. They might investigate a claim that a community committee can make 50,000 *cascarones* (eggs with confetti) in one year. Or, they might investigate the "best meal" to help families experiencing food insecurity (Suh et al., 2017; Turner et al., 2021). In modeling tasks, students must generate important quantities based on things they know or need to know. In this way, children are invited to tap into their knowledge of a situation as integral to figuring out the math problem. Mathematical modeling engages students in making sense of, analyzing, and solving tasks that are open ended and address real-world situations. Children draw on their knowledge of the context and mathematical relationships to propose solutions to modeling tasks (aka "make a model") that could answer the question. They revisit their models and adjust as necessary. Along the way, they make decisions based on their understanding of the real-world situation and their mathematical reasoning. It is another way that high cognitive demand can intersect with cultural and community funds of knowledge, making space for students to propose solutions to problems they both care about and have knowledge about (Aguirre et al., 2022).

> **What is "realia"?** *It's the use of real-world objects in the classroom, often in contrast with working with manipulatives or pictures. For example, when teaching about volume, bringing in milk and juice containers of various sizes would be a way to use realia in the classroom.*

Mathematizing real-world routines can also be a platform for high cognitive demand tasks, while taking less time to implement than a full cycle of modeling. These routines complement mathematical modeling very well because students generate their own questions from an image, a video, or other artifact from the real world (e.g., maps, realia, etc.). In this way, the cognitive demand is sustained because children not only solve genuinely problematic math problems but also problem pose (Simic-Muller et al., 2009), meaning they propose the questions they will ask based on the situation. As children learn to be mathematical problem posers, they also learn that their ideas are important and that they can be generators of mathematical questions, not just mathematical answers. (For more on this routine, see the Mathematizing the World activity box.)

Suh et al. (2017) propose a cycle for mathematical modeling that fosters 21st-century skills while tapping into children's funds of knowledge. In Figure 4.2, critical thinking, communication, collaboration, and creative problem-solving stem from children's engagement in cycles of posing a problem based on real phenomena, gathering information, making a model or first solution, then analyzing, evaluating, and revising as necessary. This mathematical modeling cycle can be a starting point for your own classroom, as you engage students in mathematizing activities.

FIGURE 4.2 Mathematical Modeling Cycle

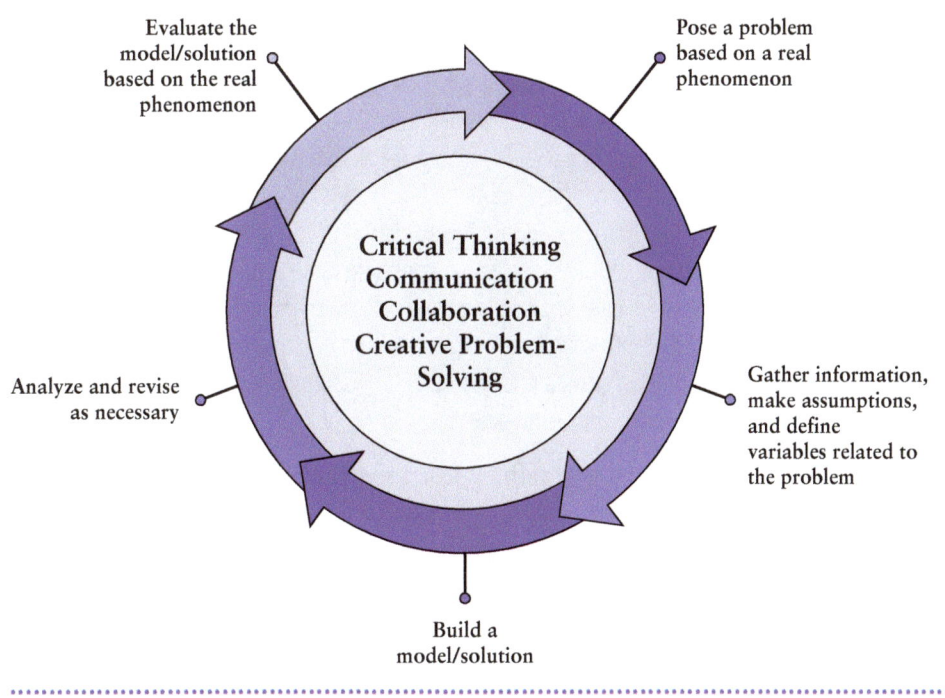

SOURCE: Suh et al. (2017).

ACTIVITY: Mathematizing the World

Mathematizing in its simplest form means seeing mathematics in everyday activities or real-world situations. There are a variety of ways that teachers can start mathematizing with their students. We encourage teachers to build on what they learned by engaging in the activity from Chapter 3 Dimension 1 and expand it to include looking for the mathematics in a situation, image, or video with students as a teaching activity:

- **Take a picture or a video from around the neighborhood and engage in problem posing.** This could be because of a significant event taking place (e.g., state fair, Ferris wheel, or swap meet), a beloved location in the community (e.g., pumpkin patch, water park, or library), a fantastical image (e.g., giant crocodile, whale, or tree), or because you were struck by something mathematically intriguing. For example, during the Black Lives Matter demonstrations in the summer of 2020, a teacher in Seattle engaged her students in mathematizing with a picture of a large crowd facing police officers. The invitation to the students was to "notice, wonder, and connect" with what was going on in the photo. The teacher then captured students' math questions on poster paper, and as a class they decided which ones to generate solutions for. With such an example, students might be curious about how many people are in the crowd. They might wonder about the ratio of people to police officers. Or, they might wonder how much area is covered by the crowd. All of these questions connect with rich mathematical concepts that can be explored together. See www.eqstemm.org for a mathematizing the world planning guide.

- **Read an illustrated book with your class and engage in mathematical noticing.** Pause as you read and invite children to tell you what they see in the pictures, what quantities they hear, and what they wonder about that can be answered with mathematics. Maria has written about mathematizing children's books as a way to engage in discussions of racism and to develop quantitative reasoning alongside students' sense of social justice. One suggestion she has is to ask the quantitative questions alongside other wonderings children have and highlight how mathematics can be used to answer some of their questions. As with the previous suggestion, teachers capture children's questions so you can explore them more deeply later. Read more at https://mariazavalaphd.com/2020/06/24/talking-race-and-math-with-kids-infusing-quantitative-lenses-into-read-alouds/

- **Analyze maps with children as a starting point for problem posing.** Maps can be a valuable resource for problem posing. Many kinds of maps can be generated from a variety of websites and even tailored to one's specific needs. Bringing a map into a classroom can be as easy as taking a screenshot from a mapping resource or as complex as building a custom data map (see for example, National Geographic's Mapmaker: https://mapmaker.nationalgeographic.org/). For example, maps from purpleair.com came in handy for problem posing with kindergartners during the wildfires in California. By looking at a map of their local area, children became curious about air quality and about what different numbers on the map meant (see Figure 4.3; Zavala, 2023). There's no limit to the work you can do with maps—pick a map and invite children to get curious about what the data on the map mean.

FIGURE 4.3 An Air Quality Map Used for Mathematizing With Young Children

SOURCE: purpleair.com

Children need ample experience and opportunity with challenging, nonroutine mathematics tasks to develop as critical thinkers, as well as to showcase their brilliance. Seda and Brown (2021) argue that high cognitive demand is also how teachers show they hold high expectations for their students: "A steady diet of low-level worksheets that place a very low cognitive demand on students produces mathematically anemic students who will never have the opportunity to learn how to persevere or make sense of problems. The types of tasks teachers assign communicate their beliefs about students' abilities more than any words could" (p. 166). As culturally responsive mathematics teachers, we remedy mathematical anemia by fostering mathematical health, through the richness and complexity of tasks.

PAUSE AND REFLECT

- How do you show your students that you have unequivocal faith that they can engage in high cognitive demand tasks?

Understanding the Rubric for Dimension 4

The rubric for Dimension 4 (Figure 4.4) prompts us to consider particular ways of engaging with mathematics that require deep thinking, thorough analysis, and multimodal communication. The essential question for *Sustaining High Cognitive Demand* is *How does my lesson enable all my students to closely explore and analyze math concepts(s), procedure(s), and problem-solving/reasoning strategies?*

At Level 1, everything is focused on the teacher and direct instruction, where students take the role of knowledge receivers. Typically, this level means there is limited or no evidence of student ideas in the lesson, which may be characterized by only the teacher talking. Children's ideas are at the margins. As the ratings progress, students have increased responsibility to do the mathematical work and communicate their ideas and strategies. The focus shifts from a select group of privileged students toward every student having the opportunity for deep mathematics engagement and analysis. Level 5 means that activities that engage all students in complex thinking and analysis of ideas, through multiple modalities, permeate the lesson. Sense-making through justification and mathematical argumentation are centered and prioritized. Multiple forms of communication and ideas remain valuable to the collective work of the classroom.

Teaching Story: Sustaining High Cognitive Demand

In a kindergarten classroom in a dual-immersion (Spanish and English) school, kindergarten children have been learning about quantitative concepts like more, less, and equal. The school has also been working on an initiative

FIGURE 4.4 Rubric for Dimension 4: Sustaining High Cognitive Demand

Fragile/
Margin

Strong/
Centered

1	2	3	4	5
Students receive, recite, or memorize facts, procedures, and definitions. There is no evidence of conceptual understanding being required. There are no opportunities for mathematical problem-solving, mathematical analysis, or exploration.	Students primarily receive, recite, or perform routine procedures without analysis or connection to underlying concepts or mathematical structure. There are some opportunities for mathematical exploration, but activities do not require analysis to complete. OR A select group of students get access to activities requiring authentic problem-solving, analysis of procedures, concepts, or underlying mathematical structure.	At least one sustained activity involves all students with complex problem-solving, analysis of procedures, concepts, or underlying mathematical structure. There is at least one sustained activity that requires mathematical exploration, analysis, and explanation.	Most of the lesson involves all students in activities that require close analysis of procedures, concepts, or underlying mathematical structure. OR involve complex mathematical thinking, use multiple representations, and demand justification.	The entire lesson involves all students in activities that require close analysis of procedures and concepts, involve complex mathematical thinking, use multiple representations, AND demand explanation and justification.

to reduce food waste. The teacher, Ms. Rosario, creates a problem around the experiences of one student, Alejandro, who brings his lunch from home and is proud of how he packs it every day. Ms. Rosario introduces the activity, a new modeling task, inviting children to figure out the following: *How much of each kind of food should Alejandro pack in his lunch?* The task is designed with specific graphic scaffolds, but aside from that, Ms. Rosario is leaving the selection of amounts, the drawing of the items, and how children total their amounts up to them. After a brief planned launch to ensure children understand the context and what the task is asking, children are asked to work on their own solutions to the problem, on their own or in consultation with their table groups. As she circulates, Ms. Rosario prompts children to select the amounts that seem reasonable and to use their experience with what makes for enough food without wasting it. She asks them to draw their answer inside the lunchbox image provided. Some students decorate the "lid" of the lunch box as well as draw their items inside. As a challenge, they are asked to write an equation that shows how they got their total. Selecting amounts and drawing them all together inside of the lunchbox is what the teacher hopes every child will be able to show for their model, and as she circulates, she prompts children to tell her if their amounts matched their drawing. The extra challenge of writing an equation revealed how children were learning to play with symbols.

In the following example, Alejandro (the student featured in the word problem) elects to pack six baby carrots, one string cheese, one apple, and two brown rice crackers. He draws amounts that match his plan, showing all ten items. He then experiments with symbols to show how he totaled up his amount. The teacher noticed that he worked to include all his numbers and some plus signs, and although he didn't write out a number sentence as we might have expected ($6 + 2 + 1 + 1 = 10$), he attempted to include the numerals and the signs, and the final number he wrote was his total. Ms. Rosario asks Alejandro to be one of the students who shares his solution (see Figure 4.5).

In a closing discussion, Ms. Rosario invites the children back to the rug and asks for some volunteers to share their work. This open-ended problem results in a variety of responses from children and many enthusiastic volunteers who want to show off their pictures as well as their calculations. As children share, Ms. Rosario asks the question "And how many items are packed all together in your lunchbox model?" Many children have a prepared answer, having totaled up their amount and written it on their paper, but some start counting the objects in the picture by ones when asked, giving another chance to engage in counting. She also asks, "How did you decide it was enough?" Children share their thoughts, ranging from "It's okay if he has leftovers because he can take them home" to "I just know it would be a good amount."

As they share, children are invited to compare their own thinking to that of others, sometimes disagreeing on what good amounts of food were. For example, some children concerned with food waste chose far less amounts

FIGURE 4.5 Alejandro's Work on the Lunchbox Problem

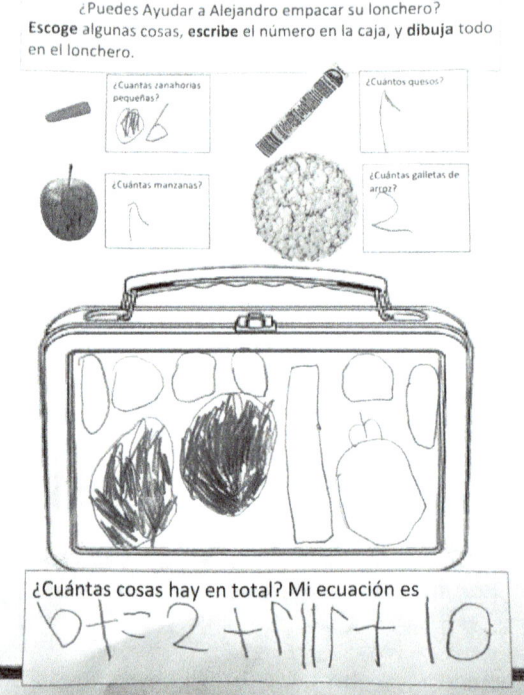

¿Puedes Ayudar a Alejandro empacar su lonchero?
Escoge algunas cosas, **escribe** el número en la caja, y **dibuja** todo en el lonchero.

¿Cuántas zanahorias pequeñas?

¿Cuantos quesos?

¿Cuántas manzanas?

¿Cuántas galletas de arroz?

¿Cuántas cosas hay en total? Mi ecuación es

The foods and amounts depicted in the image are baby carrots: 6, string cheese: 1, apple: 1, and brown rice cracker: 2.

than one student who was having fun experimenting with large numbers (adding 100 string cheeses and 100 rice cakes together to get 200 total items). As the students share, the teacher mentally notes the variation in equation writing, with some children leaving the box blank or only writing the total they counted. In this way, she knows the problem was generally cognitively demanding for her class, with high success in how children chose amounts, represented them, and counted up their totals. Important early numeracy skills were also observed, such as one-to-one correspondence and cardinality. The variation in equations was expected, and she knew this would be a stretch for some students, and something to continue to work on.

PAUSE AND REFLECT

- Using the rubric, what evidence do you see for Ms. Rosario's activity maintaining high cognitive demand? Note the evidence from the vignette, and also consider what could happen in her classroom to continue to sustain the high cognitive demand for students.

In the vignette, we note that Ms. Rosario maintains a high level of cognitive demand by first selecting an open-ended and mathematically challenging task for the students. By introducing choice through a modeling task, children had the opportunity to work at a variety of levels, including the additional challenge to all to write an equation. She maintains a high level of cognitive demand throughout the lesson by following the children's thinking and exploring how they decide what they are doing. Children analyze each other's reasoning as they share their ideas and connect their thinking with each other's mathematical strategies and understanding of food waste and hunger.

ACTIVITY: Learning to Let Go!

In our experience, teachers want to support children to be critical thinkers, mathematical modelers, and confident problem-solvers. They want to give those opportunities to their students, but they sometimes fear what will happen when they let go of control of the classroom. These fears are further exacerbated by unconscious bias when we presume particular students, because of their designated status as an "English learner" or a label of "learning disabled," or because of their racial or ethnic backgrounds, cannot do mathematics without a teacher showing them step by step what to do. Learning to let go is tricky, but making space for children to process and work on their own ideas, when they are in a space and place where they understand what is being asked and they are supported, is important. If we are to center marginalized children's thinking in the mathematics classroom, we have to make the space for it. For this activity, we invite you to first reflect on your own learning experiences, then select a task to give to students and check its cognitive demand, and finally to implement the task and see how you navigate keeping the cognitive demand high. Your reflection on how this activity goes could be a good starting point for the next activity for the dimension *Scaffolding Up*.

Part 1: Self-Reflection

We invite you to reflect on how you have benefited in your own education from being allowed to figure out something on your own or with a group of peers (meaning people who had similar knowledge of the skill or concept). We don't mean you were abandoned by a teacher; rather, we mean to address the following questions:

- What is a skill or concept you have come to learn that in part benefitted from your own time working on it?

- Did you have a sense of accomplishment that was different from other learning experiences?

- What was it like to realize you understood something new and that you had done a lot of the work yourself or with peers?

Part 2: Select a Task to Implement

For this activity, we ask you to first analyze a task that you think is or has potential to be a high cognitive demand task. We invite you to use a task from your curriculum, to select a modeling task from the EQ-STEM collection (www.eqstemm.org), or to use a resource you have had your eye on.

Use the Task Analysis Guide (Figure 4.1), which conveniently has bullet points that can serve as check boxes for you. Review the criteria for each area and evaluate your task against it.

- Do the check marks point to a high cognitive demand task? If they do not, why not? And if they do not, try selecting a different task before proceeding to the next section.

- Reflect on what makes the task a high cognitive demand task.

- Finally, plan how you will implement it with your class in ways that maintain high cognitive demand. Stay connected to your feelings of success and accomplishment when you worked through a task, and use that as a touchpoint for your planning.

Part 3: Practicing Letting Go!

Perhaps you feel like just doing the first two parts of this activity are plenty to stretch your thinking and expand your practice, but if you are ready for more, please read on.

- **Let go of control:** In your next math lesson, set a timer for five minutes and challenge yourself to not share any of your own mathematical ideas but instead to only ask questions. Focus your discussion with children on probing questions that elicit their ideas. Once the time is up, see how much longer you can keep it going. Then, take a few minutes and jot down or voice memo your thoughts about how that went for you and for the students.

- **Reflect on how it went:** Note where you wish you had more planned scaffolds and where you felt things were really humming along.

- **Make a plan:** How can you extend the time even more and give the control of the mathematics to the students? What will you try out next time? How will you ensure that you are giving that space to think to students who are usually denied that time?

DIMENSION 5: SCAFFOLDING UP

ESSENTIAL QUESTION

How does my lesson maintain high rigor with high support for all students?

Since its introduction to education in 1976, *scaffolding* has made its way into the mainstream vocabulary of teachers. Anghileri (2006) notes that "The notion of scaffolding has been used to reflect the way adult support is adjusted as the child learns and is ultimately removed when the learner can stand alone." Picture the scaffolding used in building construction that is gradually removed as the renovations are made. As Anghileri (2006) describes it, there are three levels of scaffolding in mathematics classrooms:

Level 1. *Classroom structures, routines, and the physical environment that support students engaging in mathematical activity.*

Level 2. *Teacher interactions with students, such as taking time to introduce the context and problem in a complex task or ask probing questions.*

Level 3. *Teacher guidance to expand students' ideas in formal mathematics, by listening and responding to mathematical sense making. Students expand their knowledge through processes of generalization, extrapolation, and abstraction.*

While teachers should attend to scaffolding as a process across all three levels, we also invite you to get curious about what forms of mathematical knowledge are valued in the classroom. Consider the weight of privilege we might unknowingly give to traditional Western notions of abstraction and generalization over other forms such as concrete representation and application. Instead, we look to find a more balanced approach, validating a variety of ways students expand their understanding of mathematics and respecting that children may move between forms when learning new concepts or explaining concepts to others.

Scaffolding is supposed to provide *temporary* supports into rigorous math. But in practice, attempts at scaffolding can water down the cognitive demand of a math task, as students' needs are often equated with students' inabilities to do the math (Stein et al., 1999). Scaffolding should be a tool to support student access to and engagement with mathematics, not a bunch of construction debris that blocks entry to the mathematical ideas. Therefore, this dimension is focused on maintaining the rigor of a task while providing scaffolds with multiple entry points into the work. In this way, we build on the work by equity scholars who push for maintaining high expectations for all students.

Instructional Approaches to Scaffolding Up

Shifting to higher expectations for mathematics instruction have helped put the emphasis on student thinking as a driver of instruction. However, in many cases, our own worries about math and being able to clearly answer student questions can lead to overscaffolding or shutting down student thinking during the lesson. To counter this, we advocate for flexibility in lesson design. Specifically, we focus this dimension on "planned supports" because we want to emphasize the distinction between preparation to support students to productively engage in complex math tasks and overcontrolling the mathematical thinking and engagement during those tasks.

> *"We want to emphasize the distinction between preparation to support students to productively engage in complex math tasks and overcontrolling the mathematical thinking and engagement during those tasks."*

Some of the planned supports are going to be explicitly used, while others are going to sit in our "back pockets" ready to use if needed.

Such planned supports could include specific lesson structures such as the following:

- A three-read strategy or other scaffolded launch of a complex task, where teachers carefully plan how they will preview the context and question in a task with students before students work on it (e.g., the San Francisco Unified School District has this resource: https://www.sfusdmath.org/3-read-protocol.html)

- A graphic organizer to help students organize their own thinking (see, for example, the design of the task in Figure 4.5 from Ms. Rosario's class)

- Specific realia or tools that may aid students in modeling mathematical situations

Social and Analytic Scaffolds

As we introduce increasingly complex problems to children, **Sustaining High Cognitive Demand** means planning for scaffolds that provide support to children that we may not have anticipated before. Anhalt (2014) introduced ideas of **social scaffolds** and **analytic scaffolds**. Social scaffolds are the support students need to work collaboratively through discourse, idea sharing, and collectively making sense of problems. For example, creating a classroom environment that is language-rich and centers, rather than marginalizes, multilingual learners is also a form of social scaffolding. Analytic scaffolds are the supports built into the task, such as making meaning of mathematical ideas, understanding the mathematics that could be used in a situation, and connecting across representations. As Anhalt describes it, "The general goals of analytic scaffolding are to connect the mathematics content to students' life-experiences and existing knowledge, and to encourage students to draw on a range of resources to communicate the meaning of the mathematics" (p. 113). In addition, all students can be encouraged to draw on a variety of communication methods (drawings, gestures, first languages, concrete objects, etc.) to be able to collaborate around the task. Anhalt argues that both types of scaffolds are interrelated and overlap. The following table (Figure 4.6) shows examples of social, analytic, and intersections of both types of scaffolding.

Back-Pocket Questions for Scaffolding Complex Tasks

Teachers can also differentiate for themselves between the supports to be built into the lesson as a scaffold for all learners and the specific "back pocket" supports to have at the ready, if needed. Teachers use their professional judgment to decide when they should be deployed. For example, depending on the situation, a teacher might choose to explicitly include a graphic organizer for a problem or may first ask children to solve it however they like

FIGURE 4.6 Examples of Social and Analytic Scaffolding

SOCIAL SCAFFOLDING (STUDENTS INTERACT AND WORK TOGETHER TO LEARN FROM ONE ANOTHER)	BOTH SOCIAL AND ANALYTIC SCAFFOLDING	ANALYTIC SCAFFOLDING (STUDENTS MAKE PROGRESS TOWARD THE MATHEMATICAL OBJECTIVE)
Collaborative Teams – Attention to linguistic proficiency – Change team members often **Explicit Discourse Patterns** – "Think-pair-share" – Sentence frames to make claims: "I think ___ because ___" – Examples of questions to probe each other's thinking: "Why do you think ___ and ___ are related?" **Establish Communication Norms** – Teams share assumptions and solutions with class – We ask questions that elicit explanations – We rephrase each other's ideas – Encourage multiple forms of communication, including home/first languages	**Create Language-Rich Learning Environments:** reference charts and posters of student work around the room; complex problem-solving and mathematical modeling are a normal part of math instruction **Mathematics Tasks Set in Contexts That Are Relevant to Students:** activities highlight a variety of children's life experiences over the course of the school year **Establish Mathematical Norms** of explaining, justifying, summarizing, questioning, and sharing solutions among students	**Explicit Connections** – Elicit prior knowledge of the context – Discuss possible relevant mathematics – Connect language to mathematical representations **Experimentation** – Encourage experimentation, and expect it to sometimes lead to incorrect solutions – Encourage varying assumptions in modeling problems – Encourage variations in "If … then … " questions **Sustained Focus** – Students understanding of, reasoning about, and reasonableness of outcomes **Multiple Representations** – Pictures, diagrams, tables or graphs – Mathematical notification and symbols – Concrete manipulatives – Graphic organizers – Essential terms and vocabulary on poster paper for reference

SOURCE: Adapted from Anhalt (2014).

on a blank piece of paper, reserving a graphic organizer if necessary. As we complete tasks that we plan to do with children, we gain new insight into the possibilities and the potential pitfalls. Anticipating students' solutions to problems can help us prepare a number of scaffolds to hold in reserve and decide when and how to bring them out.

In their work with early elementary teaching doing culturally responsive mathematical modeling, the EQ-STEMM Project leaders found that certain questions were particularly helpful to scaffold students through different phases of the modeling process. They collected examples of these questions and provided them to teachers in their study, finding that they filled a need to know how to probe students' thinking in ways that lead to renewed, productive engagement with the problem. They organized the scaffolding questions into dilemma categories. For example, Figure 4.7 shows questions they recommend to scaffold students' engagement with modeling problems while students are working through the problem.

FIGURE 4.7 Suggested Teachers' Questions That Scaffold Up

LESSON PHASE	WHAT CAN WE DO?	HOW MIGHT IT SOUND?
Small Group/ Partner Work Time	Encourage students to try out their ideas, and offer tools to represent their ideas.	• You said each person gets 3. Can you show that with number? Pictures? Cubes? • Here is a picture of _____. How can you use this picture to show your idea? • Can these cubes help you test that idea out? Give it a try and let me know how it goes. • You are on the right track. What do you need to figure out next?
	Support students to clarify and label their models/solutions.	• What does this number 15 over here represent? • Can you add words to your drawing to clarify the parts? • How does this number over here, this big number of 10,000, relate to your solution?
Whole-Class Discussion	Invite students to share ideas in multiple ways, from pictures, words, gestures, and multiple languages.	• Can you show us your picture? • Can you point out the key parts of your model/ solution? • You can share your idea in Khmer, go ahead. We'll listen. • Can you and your partner work together to explain your model? • Can you give us an example of how your model works?

SOURCE: Adapted from Turner et al. (2019); www.eqstemm.org

By keeping the back-pocket questions on reserve, you also give yourself the space to learn about the struggles the students are having before overloading them with ideas. In this space between setting up the task and providing extra scaffolds, you have time to navigate the decision you will make to ensure that the students are still getting the chance to think and try out ideas, while also being ready to support students who are genuinely in need of additional support.

Ultimately, *Scaffolding Up* is about sustaining the cognitive demand of a task while providing all learners access to the task. Culturally responsive teachers approach scaffolding by first presuming children are competent problem-solvers and then using their understanding of children's personal strengths and struggles to plan for appropriate support. We must lean into the tensions between providing enough scaffolding to students to access the task and trusting in children to do the work of sense-making and problem-solving. We expect that we won't always get the balance right. Children may draw on personal and collective knowledge in ways that surprise us, and at other times we will reflect on how we could have supplied more support.

> *"Culturally responsive teachers approach scaffolding by first presuming children are competent problem-solvers and then using their understanding of children's personal strengths and struggles to plan for appropriate support."*

From Scaffolds to Tools for Thinking: Repurposing Manipulatives

As we mentioned earlier, typically scaffolding is meant to be removed at some point. As we support students' foray into the mathematics, we can assess how the students are progressing in their own abilities to access the tasks and how they move from using provided scaffolds, to taking control of their own learning and applying their own learned supports. However, we have two comments on this point. First, we advocate for a broad perspective on what it means to remove scaffolding. Traditional notions of scaffolding are steeped in our dominant culture of individualism, which privileges the individual as the unit of measure for learning. Other perspectives, such as those held by some Indigenous communities, hold that knowledge is distributed and in the extensions of ourselves in others, a collective knowing (Ullrich, 2019). Social learning theories also value the notions of "distributed knowledge" and highlight that use of language mediates learning (Vygotsky, 1978). Therefore, in CRMT we take a nuanced perspective of the purpose and presence of scaffolds. Collaboration is a kind of scaffold, but collaboration is not a means to an end. The act of working in groups can lead to powerful learning, yet we wouldn't say that the group is there to scaffold the learning of only the individual. Groupwork is generative and powerful unto itself. In Chapter 5 we will talk more about collaboration.

So there are some techniques that are traditionally thought of as scaffolding that we would suggest repurposing to more general classroom teaching strategies. That is to say, some supports are useful not just as a tool to get into the mathematics but also as a tool to continue and model one's own thinking. Base ten blocks, for example, are a manipulative that at times may serve as an explicit scaffold but should always be made available to learners as a tool for both problem-solving and modeling to others how mathematical ideas work. In that sense, we don't advocate for removing access to manipulatives. Removing the manipulatives suggests that they are only useful to scaffold learning and that real mathematics takes place on the paper (or in your head, etc.). Rather, we recommend fostering an environment where manipulatives and other concrete objects are readily available and used as both a means to learn and a means to communicate ideas. As students learn to make their own choices about how to dig into the mathematics and present strategies using different representations, they are also engaging in a core mathematics practice: using appropriate tools strategically. In the case of using blocks and other concrete manipulatives, the strategic thinking is both about how to solve the math problem at hand and how to communicate and collaborate with others.

Understanding the Rubric for Dimension 5

Engaging this dimension requires teachers to have a clear understanding of the cognitive demand in the activities of the lesson. Then, planning for the lesson draws on engaging scaffolds for the whole class, and subgroups of students in particular. The essential question for Dimension 5 is *How does my lesson maintain high rigor with high support for all students?* Figure 4.8 shows the rubric for Dimension 5.

A Level 1 would either be the absence of planned supports or diminishing cognitive demand disguised as a scaffold. For example, if a teacher indicates that they plan to provide some students with alternative work if this problem is "too hard," we would call that a Level 1. At Level 2, there may be scaffolding that puts too much control in the hands of the teacher or scaffolds only at the beginning of the lesson. This may be evident in a lesson plan when a teacher has planned a very long launch or plans to first explicitly model a solution strategy before introducing a math task. Moving up to Level 3 requires planning supports that maintain cognitive demand but may not permeate the whole lesson. Level 4 is when the supports are appropriate and take into consideration subgroups or individual students. Level 4 also includes drawing on known student strengths in parts of the lessons. Full realization of this dimension (Level 5) would include multiple planned supports for all phases of the lesson that attend to most students and draw on students' known strengths.

Teaching Story: Scaffolding Up

It is springtime in Ms. D's third-grade class, and Ms. D is setting up to do a modeling task with her students. She puts up an image of brightly colored eggs. Some eggs are broken with paper confetti spilling out. Ms. D asks, "Has anyone seen these kinds of eggs?" One student shouts out, "At my

FIGURE 4.8 Rubric for Dimension 4: Sustaining High Cognitive Demand

	Fragile/ Margin				Strong/ Centered
	1	**2**	**3**	**4**	**5**
	There is no evidence that the teacher has planned supports in ways that maintain the rigor of the task while providing access for students.	Planned supports provide too much scaffolding and diminish the rigor of the task. OR Planned supports may only attend to access at the start of the task, not throughout the lesson.	Planned supports maintain rigor but may not connect to either this *specific mathematics task,* or draw on the strengths of students. Planned supports may only attend to access at the start of the task, not throughout the lesson.	Specific planned supports ensure most of the class understands the task and have a way to get started. Planned supports are used throughout the lesson, although planned supports for individuals or subgroups may not directly connect to known student strengths. There is no evidence or minimal evidence of supports planned or used for individuals or subgroups of students.	Specific, planned supports address the whole class, as well as individual or subgroup needs. Planned supports are used throughout all phases of the lesson, including launch, student work time, strategy sharing, or lesson wrapup. Planned and enacted supports for subgroups differ from those for the whole class and build from students' known strengths.

cousin's friend's birthday party." "What are they called?" asked Ms. D. The child responds, "cascarones; it means eggs with confetti." Everyone says out loud, "*cascarones*." Ms. D thanks the child for translating the word and repeats "cascarones." She continues the conversation by asking, "What do you notice about this image?" Students raise their hands and respond excitedly. "I notice different colors." "I notice they are in arrays." "I think each one has fifteen." Ms. D writes their responses down on the poster paper. Next, Ms. D shows a picture of a group sitting at a table making the eggs. There is text next to the image. Ms. D reads, "The members of la fiesta committee say they can make 50,000 cascarones in one year. Is this possible? Turn and talk to your partner about this situation." As Ms. D listens to students, she hears statements like "I think no, maybe 50," "you have 12 months right?," and "that's a lot of eggs with confetti." Ms. D gathers the group back. She asks, "Okay, everyone, what do we need to know to figure this out?" Hands shoot up in the air. Oscar says, "We need to assume about 100 people." Ms. D clarifies, "like 100 workers? Okay, Oscar says we need about 100 workers. I'm not going to put an exact number here for workers, because some of you might have different ideas. But we need to know the number of workers on the committee. Okay, what else?" Abigail raises her hand and says, "I think they can probably do 1,000 a day so they need to make 500." Ms. D revoices, "So you think 1,000 eggs; we need to assume a certain number of eggs per day. Okay, I'm not going to write a specific number down; your group gets to decide." The conversation continues as Ms. D writes other ideas down on the board. She then states, "I want you to think about your own parents. How often do they work? Does that matter?" The kids shout out what they know about their own experience. Michelle says, "My mom works every day, including Saturdays and Sundays." Ricky says, "My dad only works summers." Then Elise says, "It depends on if they work four hours a day or more." And Melissa offers, "What if they never close? They work 24 hours?" Ms. D revoices, "So it sounds like something else that is important is how many days a week does the committee work and how many hours in each day?" "Yeah!" shout the kids. Ms. D says, "Okay, I think you are up for the challenge. You are to work in your groups to answer these questions and determine if the committee can make 50,000 cascarones."

As students continue to work on the task, Ms. D walks over to a group discussing the number of workers and how long they work. After listening to the group, Ms. D summarizes, "So, you say that you have 50 workers and they are working 15 hours a day. Is that reasonable?" The students, respond, "Yes, they have to sleep eight hours and work the rest of the time." Ms. D asks them to show their math thinking to help explain the reasonableness of their ideas, especially if working fifteen-hour days is reasonable. She moves to a different group with two boys working together, Alex and Kenyan. They proudly state, "The workers make 500 eggs per day."

Ms. D asks, "How many is that per week?" The boys reply in unison, "2,500." "How do you know? What does the five represent in this equation?" Alex explains that a week is really five days. Ms. D then asks them to consider the original claim again. "If you want to see if they make 50,000 eggs in one year, what do you need to figure out next?" The students talk quietly and then respond, "We know—twelve months." Ms. D affirms their thinking, "Yes, you are on the right track. What do you need to figure out next?" One child responds, "2,500 × 4." Ms. D asks, "Why 4?" Kenyan replies, "Because there are four weeks in a month." Ms. D. says, "Oh, that is a fact you know already. Perhaps that will help you." And she moves to a different group.

During group presentations, Alex and Kenyan present their solution. They describe how they assumed 100 workers would make 500 eggs. They indicate that the workers "would do five days a week and eight hours, so they don't only have to go home and start sleeping and then wake up and have to go to work. So they can spend time with their family, too." Ms. D affirms the idea that their solution took into consideration the workers making time for family. Alex then explains that the workers would make 10,000 eggs in one month: $4 \times 2,500 = 10,000$. The teacher confirms that their response is 10,000 eggs in one month, and then asks them to respond to the original question, "What did you find out?" The boys state, "Yes, they can make 50,000 cascarones in one year." The teacher looks at the class and asks, "Any questions or compliments?" One student asks if they could explain what the 4 meant in their equation. Alex explains, "Because there's four weeks in one month, and they're doing 500 a day. So, it would be four weeks times 2,500. And so, it'd be 2,500, because it'd be $4 \times 2,000$ would be 8,000. And then 4×500 so that'd be 2,000. And that would be 8,000 plus 2,000 which would be 10,000. And if after 5 months of a year, they'll already have 50,000." Ms. D, responded to the student who asked the question, "Does that make sense to you?" The student nods affirmatively. Ms. D. thanks the presenters and the class for their hard work and perseverance.

In the vignette, we see Ms. D drawing on a number of scaffolds to move students' thinking along, while maintaining high cognitive demand. From the start, her launch supports students to understand the context and the problem they are solving. They are invited to make assumptions and question the reasonableness of those assumptions. She asks a number of probing questions as she circulates, scaffolding students into clarifying the models and making their mathematical thinking visible. As she brings the class together to discuss at the end, she draws on another set of talk moves to bring the class along with the pair of students presenting their ideas, including returning to the original question and probing students' thinking about the reasonableness of their assumptions.

ACTIVITY: Building Capacity to Balance Cognitive Demand and Scaffolding

In this chapter we introduced some approaches to scaffolding, including attending to social scaffolding and analytic scaffolding.

Take a few minutes to revisit the Scaffolds Table in Figure 4.6. We invite you to make a copy of the page and mark it up as you consider the following:

- What scaffolds are familiar to you? When you have implemented them, how have they supported children to engage in mathematics?

- In Ms. D's cascarones lesson, what scaffolds do you see at work? Which are social scaffolds, and which are analytic scaffolds?

- As you reflect on the scaffolds suggested in this section, how could you imagine implementing them in the future?

DIMENSION 6: AFFIRMING MULTILINGUALISM

ESSENTIAL QUESTION

How does my lesson make space for multilingual learners (MLL) to be central participants in mathematics activities?

In some parts of the United States, increased attention on how to support English learners (who we call multilingual learners or MLLs) has resulted in a growing research base of what teachers should do to support both learning mathematics and English language development. However, this research has a variety of assumptions about what meaningful engagement with mathematics is and what constitutes so-called "best practices." Culturally responsive mathematics teachers draw on the literature base that emphasizes the sense-making and linguistic funds of knowledge that MLLs bring to the classroom. As Moschkovich (2013) has shown across her research, the language of mathematics is far more complex than vocabulary—it is "the communicative competence necessary and sufficient for competent participation in mathematical discourse practices" (Moschkovich, 2013, p. 46). Therefore, it is important for teachers to know the histories of schooling, languages, and learning mathematics that MLLs bring to the classroom. In particular, Moschkovich emphasizes that all languages students bring to the classroom are a resource for mathematics learning. From a sociocultural learning theory perspective, we could understand this claim by considering how language mediates thought and action (Vygotsky, 1986). Full access to all linguistic resources will increase sense-making of the mathematics, while reducing access to linguistic funds of knowledge by, for instance, requiring students to engage in English only or prioritizing the completion of work in English only, limits learning opportunities.

When we talk about **Affirming Multilingualism**, we are talking about affirming children. Language is identity. It is not possible to disentangle who students are from the languages they use to make sense of the world, be a part of their family, and communicate with others. At the same time, we also have to acknowledge the power dynamics around English and all other languages in the United States.

> *"When we talk about Affirming Multilingualism, we are talking about affirming children. Language is identity. It is not possible to disentangle who students are from the languages they use to make sense of the world, be a part of their family, and communicate with others."*

Globally, the English language holds incredible power (Macedo et al., 2003). For example, it may come as a surprise that in bilingual classrooms, such as Spanish immersion settings, where mathematics is taught in Spanish, Spanish speakers don't necessarily outperform their English-dominant peers. In work with preservice teachers in dual-immersion settings, research has shown that white students learning Spanish may still have enough resources in mathematics to be more central participants than Latinx students who are fluent Spanish speakers and that this dynamic is hard to disrupt without specific attention from teachers (Zavala, 2017). Even when the language of instruction shifts from English to another, power imbalances permeate the classroom without explicit attention to how multilingualism can be moved from the margin to the center.

Affirming Multilingualism is about using all the resources, even if you as a teacher don't have every resource, to reach out to families and communities. When the classroom is genuinely multilingual, when the English-other binary has been disrupted, there is space for multiple forms of communication to be used and valued.

Recent research on the concept of *translanguaging* sheds light on the fluid and holistic ways people engage their linguistic resources as they learn mathematics (Maldonado Rodriguez et al., 2020). Translanguaging is the process of fluidly using all of one's linguistic resources (multilingualism, reading contextual clues, empathy, etc.) to engage in

> *"Translanguaging is the process of fluidly using all of one's linguistic resources (multilingualism, reading contextual clues, empathy, etc.) to engage in sense-making and discourse."*

sense-making and discourse. We have long been familiar with ideas of code-switching and concerns as to whether students have a good enough grounding in the official language of school to be competent readers and mathematicians. Translanguaging is an evolution of code-switching, embracing the intersection of linguistic resources and recognizing that within the overlap is a linguistic resource unto itself—a *trans*language that crosses borders.

Critiques of translanguaging usually follow along the lines of watering-down official languages, such as engaging in "Spanglish" if you aren't fluent in one or the other. However, as Ofelia Garcia (2009) asserts, attending to just the competencies students have in so-called "official" languages is not a good picture of their actual linguistic resources, which span multiple languages. She contrasts *translanguaging*, the process of using all of one's linguistic repertoire to engage in meaning-making, with traditionally "legitimate" language practices, which are based on the command of officially recognized languages often defined by countries and marked by their borders. In this way, she positions translanguaging as also having political power, as it potentially challenges the legitimacy of borders (in particular, borders between Mexico and the United States, and other borders that students and families may symbolically or physically cross again and again as a regular part of their lives). In Chapter 8, Spanish bilingual teacher Melissa Adams Corral expounds on how translanguaging is central to how she cultivates a culturally responsive mathematics classroom praxis where multilingual children are centered as mathematical people.

Instructional Approaches to Affirming Multilingual Students in the Classroom

In general, approaches to *Affirming Multilingualism* in the classroom start from cultivating a language-rich environment and inviting students to be central participants using their full linguistic repertoires. Designing a discourse-rich learning environment is the starting point for positioning multilingual learners. Without ample opportunity for discourse, students will not engage in discourse. The mathematics classroom must be a place where students are invited to work on collaboration-worthy tasks, draw on their own mathematical ideas, work on cognitively demanding tasks, and draw on norms to engage with each other's ideas. Creating a discourse-rich classroom means that now you have the opportunity to position learners within the discourse community you have created and can start using explicit strategies to position multilingual learners as leaders and strong mathematical thinkers within the class.

Chval et al. (2021) advocate for keeping certain big ideas in mind, some of which we summarize below. We recommend their book, *Teaching Math to Multilingual Students: Positioning English Learners for Success (K–8),* as a resource to cover these ideas in-depth:

- **Focus on positioning:** We must focus our efforts not just on supporting and helping multilingual students to access the mathematics but also on positioning them as competent mathematicians and contributors of mathematical knowledge. Teachers must use their role in the classroom to position multilingual learners as central participants, as leaders. Chval et al. (2021) also remind us that even when we do nothing, we are still positioning students: "It is not a matter of if a teacher positions students, but how the teacher positions" (p. 13). Focusing on how we are positioning multilingual students draws attention to our actions and allows us to work with intention.

- **Situate mathematical tasks in culturally relevant contexts:** This concept is heavily anchored in Dimension 1 of the Culturally Responsive Mathematics Teaching Tool (CRMT2) but bears repeating. Research has shown that there is a connection between situating mathematics tasks in culturally relevant contexts and the way that multilingual students can access those tasks (Dominguez et al., 2014). Technology can be used to bridge the complexities of language differences; for example, a video can be shown instead of a teacher attempting to explain a situation. The combination of imagery, gesture, and spoken word can help convey understanding of a context that can then be leveraged for mathematical learning.

- **Facilitate effective partnerships between peers and multilingual learners:** It is important to pair students with peers who speak the same language, but at other times we have to consider other aspects of peer partnerships. Strategies for establishing peer partnerships include implementing social scaffolds as described in Figure 4.6, attending to how pairs work together and revisiting partnership norms frequently with students. Norms such as "respectfully disagree" and "explain your solution to a partner" may need additional support like sentence frames and opportunities to practice with nonmathematical examples.

- **Use visuals, gestures, and other scaffolding techniques:** Teachers may be well aware of using visuals, gestures, realia, and anchor charts in their teaching as ways to connect contexts and situations to students, but Chval et al. (2021) argue that these techniques also help students make sense of mathematical vocabulary, concepts, procedures, and models. One way to do this is to embed the teaching of new vocabulary within content. Rather than preview vocabulary by writing definitions, students can learn what particular words mean in the context of learning about how they are used. For example, you might invite children to solve an equal-sharing problem such as 2 children sharing 5 cookies, and discuss how we know what to call the little pieces we make when we split up cookies. This introduces the need for a word like *fraction*, and students can learn the formal word connected to ideas of "pieces," "split up wholes," and "parts I make" through the task of equal-sharing.

- **Take time to explicitly make sense of how language works:** Children who are multilingual and learning mathematics in English may come across words that sound similar to other words in English but hold special meaning in the discipline of mathematics. Linguists call this the mathematical "register." For example, *remainder* and *reminder* sound very similar, but *remainder* is the word we use for what's left over in division and *reminder* doesn't hold a special mathematical meaning. Some words are exactly the same but have unique meaning in the mathematical register, like *formula* (which doesn't have connections to baby food in the math register, but students may note they have heard of baby formula before). Taking the time to acknowledge the connections children make between sounds of words and their specialized meanings, as well as pausing to help children notice connections, will aid in students learning academic language in English.

- **Encourage writing in mathematics:** As a society, we usually don't emphasize writing enough in mathematics classes, almost seeing it as a break from writing. But writing is an important vehicle for communicating mathematical ideas. When multilingual students are encouraged to write and improve on the writing of their mathematical ideas, they not only learn concepts more deeply but also have another vehicle to being recognized as mathematically proficient.

- **Involve parents and families:** "Culturally sustaining practices that enhance their participation and engagement at schools—particularly in mathematics classes—are required since parental engagement is positively related to children's achievement in school" (Civil & Menéndez, 2010, p. 181). It is important that teachers engage parents as parents. First, you must learn from parents what their beliefs and ideas about education are and ways that they educate their children at home. Invite parents to share what they expect of school and you as a teacher as well. Parent-teacher conferences can be a good time to discuss with parents their expectations as well as answer their questions. Information you learn from parents can also be the basis for designing culturally relevant mathematics tasks.

Acknowledging the Tension of Valuing Multilingualism and Teaching English While Teaching Math

Ultimately, this dimension challenges an English-centric position while also acknowledging the need to aid multilingual students in developing their English language competencies. Thus, in the rubric, we include the use of *English as a Second Language* or *ESL* strategies, which support access to mathematics as well as connecting the language of instruction and students' home languages (their L1). We acknowledge this as a tension in the work of culturally responsive mathematics teachers. On the one hand, we work to value the translanguaging and multilingual practices that children bring to the classroom, in the name of making sense of mathematics. But on the other hand, we acknowledge the realities of schooling in the United States where there are expectations that children are supported to learn English while they learn mathematics. In fact, parents might also tell you that this is what they wish for their child, to learn English from school. Hence, the new reality of multilingualism, the concurrent development of many languages, is what we center. Culturally responsive teachers foster the multilingual classroom using strategies and resources we described above.

BUT ISN'T EVERYONE A LANGUAGE LEARNER?

Some early elementary teachers love to say that everyone in their classroom is a language learner. While we understand the sentiment of the statement, it's more accurate to say throughout life, all of us are usually learning new languages that peers and community members introduce us to. However, most of us have language competencies that come from home, even children in early elementary classrooms. So, while it is true that young children continue to learn language, they also have a wealth of linguistic resources at the ready as they enter kindergarten. Further, the statement "Everyone is a language learner in my classroom" doesn't account for the power dynamics around English versus all other languages, or what Macedo et al. (2003) call the hegemony of English. They argue that English has been given unearned superiority status, as one can see in how the English language is tied to the mythologies surrounding so-called true American ways of being. This power differential also plays out in critiques of Black Vernacular English (or Black English), when teachers insist that their students learn the "correct" way to express their ideas in English instead of valuing the linguistic resources students are using to express their ideas already. Instead of claiming that all children are language learners, we should acknowledge that all children build on their linguistic resources to continue to learn new languages. As teachers we have to work to disrupt the entrenched ideas of whose language is more correct. In the mathematics classroom, we need to value multiple forms of English alongside multiple languages.

Understanding the Rubric for Dimension 6

The rubric for Dimension 6 (Figure 4.9) calls attention to how multilingual students are positioned in relation to the mathematics. First, there should be explicit acknowledgment and attention to their own linguistic funds of knowledge. All language resources should be seen as a benefit, not a deficit. The essential question for Dimension 6 is *How does my lesson make space for multilingual learners (MLL) to be central participants in mathematics activities?*

At Level 1, a teacher ignores linguistic resources or isolates the students who are in the process of learning English, positioning them as less worthy of doing mathematics than their peers and as having nothing to contribute to their peers' learning. With Level 2, there is some acknowledgment of multilingualism, perhaps just in a lesson plan but not evident in instruction, and a teacher may be beginning to make sense of how this all works in the classroom through tolerating languages other than English but may still focus attention on MLLs solely to correct their English. At Level 3, there is increased attention and explicit connection during instruction to MLLs' linguistic funds of knowledge, and even if a teacher does not speak

FIGURE 4.9 Rubric for Dimension 6: Affirming Multilingualism

Fragile/Margin				Strong/Centered
1	2	3	4	5
There is no acknowledgment of MLLs' linguistic funds of knowledge. MLLs who are not yet fully proficient in English are ignored and/or seated apart from their classmates.	There is acknowledgment of MLLs' linguistic funds of knowledge, but they are not leveraged in lesson design. Students' use of L1 is tolerated. Teaching focuses on correct usage of English vocabulary only. No explicit attention is paid to scaffolding access for MLLs.	There is at least one instance of attention to MLLs' linguistic funds of knowledge that is central to the lesson, such as encouraging translanguaging. Even if a teacher does not use L1, it is evident that MLLs' linguistic repertoires are valued and that they are encouraged to build on them (e.g., students can present in L1, students work in groups in L1). There is at least one instance in which an English as a Second Language (ESL) scaffolding strategy is used to develop academic language (i.e., revoicing, use of graphic organizers, activation of prior knowledge, and strategic grouping with bilingual students).	Clear attention is paid to MLLs' linguistic funds of knowledge throughout the lesson. Focus is on mathematical discourse in L1 and English, not students' production of "correct" English. There is sustained use of at least two ESL scaffolding strategies, such as the use of revoicing and attention to cognates, direct modeling of vocabulary, strategic grouping with bilingual students, use of realia, graphic organizers, or encouragement of L1 usage is observed at least between teacher and one student or small group of students. The focus is on positioning of multilingual students as central participants through recognizing their mathematical competence.	Extensive and sustained attention is paid to MLLs' linguistic funds of knowledge throughout lesson. Sustained encouragement of L1 usage, or hybrid language (e.g., code-switching) is observed between teacher and students and among students, in a variety of interactions (teacher-students, pair, small group, and whole class). The main focus is the development of mathematical discourse and meaning making in both L1 and English. Deliberate and continuous use of multiple ESL strategies, such as gesturing, use of realia, use of cognates, revoicing, graphic organizers, and manipulatives are observed during whole class and /or small group instruction and discussions. The main focus is the development of mathematical discourse, identity, and meaning making as learners are positioned as mathematically competent leaders and thinkers.

all the languages of her students there is obvious encouragement and support to use other languages in the classroom. At Levels 4 and 5, teachers engage MLLs in mathematics discourse practices, with sustained attention to linguistic funds of knowledge and multiple explicit supports to affirm the linguistic AND math identities of students. In particular in Level 5, multiple forms of hybrid discourse practices (using multiple modalities, languages, and forms of communication) are happening between students and teacher.

Teaching Story: Affirming Multilingualism

This story is adapted from the marshmallow video from the Annenberg Learning Mathematics Teaching Library.

Mrs. Torrejón is launching a mathematics task with her second graders, who are preparing to go on a school-sponsored camping trip. They are almost all from Spanish-speaking homes, and many families send their children to this bilingual school so that they develop both Spanish and English language skills while learning other academic contexts. They had a homework assignment to ask their parents how many marshmallows were reasonable for one child to eat. She passes out a sticky-note to each child and asks them to write down the number they discussed with their parents. She asks them, "Who thinks they have the highest number? What about the lowest number?"

She then gestures toward a large piece of poster paper, which is blank. She asks the students, "Is our graph ready to go?" There is a chorus of "nos." She asks them questions about what they need to draw to set up a bar graph, and once they have their bar graph established, she invites children to put their sticky notes in the right spot that matches their number. Together, they construct a large visual display of their data, for each child to look at and wonder about. This is not their first time working with bar graphs, so Mrs. Torrejón is confident they can do it with minimal help. She lets the students place the sticky notes, and she lets them self-correct if they want to stack them slightly more neatly or leave them as they are.

Mrs. Torrejón does a lot of this kind of decision-making. She believes that children need a sense of community and ownership over their classroom, and that includes the mathematical activities. She has cultivated this since the start of the school year. She is conscious not to adjust students' work for this reason.

Next, they discuss students' observations of the graph. Then she brings out the question that will set up the big math exploration: *If you have to go to the store right now and buy the marshmallows, how many do you think most are going to eat?*

A few students answer 6. This is not surprising because 6 is the top of the curve in the data distribution. But Mrs. Torrejón is slightly surprised when the students answer that they chose 6 not for mathematical reasons but because of their life experiences. Marisa says, "6 makes sense because they shrink when you roast them." After noticing this makes a lot of sense, Mrs. Torrejón asks if she has another reason. She says no. Next, Anna says, "6 is good because you won't waste money." Mrs. Torrejón revoices and affirms the thinking: "Ah, so you are saying that with 6 we know they will not go to waste." She honors their thinking and does not press for a mathematical explanation that stems from the bar graph—in part because the heavy mathematical work is to come and because they provided sound reasoning from their lived experience.

She shows the class an unopened plastic bag of marshmallows. "Your job right now is to find out, if everyone eats six marshmallows, how many people can one bag feed?"

There is palpable excitement. Every group is going to get a bag of marshmallows to use. Treats like this are usually reserved for class celebrations, but Mrs. Torrejón decided that it was far more enjoyable to have the realia (i.e., the real thing) right there, so students could both play with, use to count, and then eat marshmallows.

As students work in small groups, she approaches one small group of students who have displayed their strategy on top of a large piece of poster paper. They have arranged their marshmallows in groups of 6. They have traced some of the groups with marker. Earlier, group member Erica led her tablemates to count how many groups of 6 marshmallows they had. As she moved her hand over the groups, she counted by ones in English. "One, two, three, four, five, six, seven, eight, nine, ten, eleven, twelve, thirteen, fourteen." Erica took initiative and, though she was not prompted to, she wrote down a sentence to capture her thinking (see Figure 4.10).

Mrs. Torrejón comes over to see what they have been doing. She asks Erica to read the sentence she wrote. She reads, "Hay catorce bonbones en cada grupo para catorce personas en un palito y luego los cocimos en la lumbre."

Mrs. Torrejón admires the sentence, praising Erica for her initiative and group for drawing and labeling on their poster board. She comments that it makes their thinking very clear. Then she asks the group, "So, what did you decide? How many can you feed?"

Marisa, Erica's tablemate, says, "14." And Erica replies, "No, we have more."

Marisa looks at Erica for a moment, seemingly confused. Mrs. Torrejón says, "I think Erica says there is something you should do with the 4 leftovers." Marisa lights up. "Oh, it's like we can give that to a person who wants to eat less. Yeah, so then we can feed 15."

"Very interesting. Let's share that one with the class when we circle up, okay?" says Mrs. Torrejón.

Marisa nods. Mrs. Torrejón decides it's time for the whole class to share together and starts calling everyone to the rug.

FIGURE 4.10 Erica's Group Poster

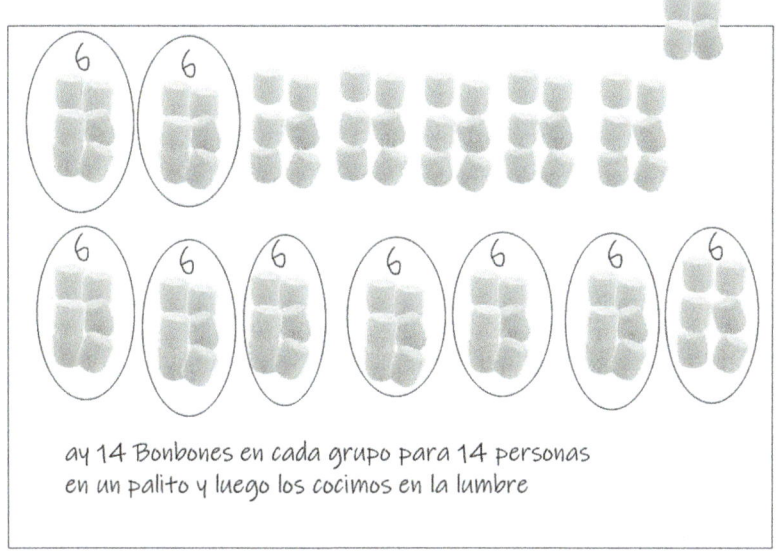

ay 14 Bonbones en cada grupo para 14 personas
en un palito y luego los cocimos en la lumbre

Erica's group has made a poster showing their solution, and Erica wrote a sentence in Spanish to go with it.

SOURCE: Marshmallow image by subjug/istock.com

Let's unpack this vignette together, since we chose to keep it long to highlight a variety of ways the teacher is *Affirming Multilingualism*. First, multilingual students are central to the mathematical activity and decision-making in her classroom. From drawing on the expertise of the home (e.g., talk with your parents about how many marshmallows are reasonable) to giving children ownership of building the bar graph, Mrs. Torrejón has multiple ways to affirm their multilingual identities and position them as competent and knowledge-able students. Children in her classroom use drawings, pictures, words, and even full-on written sentences to communicate their understanding. Erica's written work shows that multilingualism is also a regular part of the classroom, as she chose to write in Spanish and read aloud in Spanish even while most of the rest of her class participation was in English. What else do you notice about the ways Mrs. Torrejón affirms multilingualism in her classroom?

ACTIVITY: Taking Stock of How You Center Multilingualism

There is a common misunderstanding that multilingualism doesn't contribute to English learning, leading to punitive "English-only" school policies—even unofficial English-only policies in well-meaning teacher's classrooms. But recent research shows this is not the case (Steele et al., 2017). From the impact on brain physiology to the social benefits, multilingualism is not a detriment but a strength, contributing to stronger academic outcomes over time for children who are supported to value themselves and their linguistic repertoires. Whether you yourself are multilingual or not, your mathematics classroom can be a place that welcomes and centers multilingualism. For this activity, we invite you to examine your own beliefs around language and the role of school. Spend quiet time in your classroom reflecting on the following:

The Environment: How is your classroom set up to communicate that you value multilingualism?

> *What books are present? What is on the walls? How are children invited to share their thinking? You might even ask children if they feel like they can or should speak languages other than English in the classroom.*

Communication With Parents: How do you communicate to families that you welcome multilingualism?

> *When and how do you communicate with families? How do you make explicit that their languages, the home languages of the children in the classroom, are valued and an important part of the classroom learning?*

Instructional Routines and Activities: How do your mathematics routines, tasks, and homework help you communicate that you center multilingualism?

> *Is multilingualism reserved just for other subjects, or do you also incorporate it into your mathematics tasks?*

Share your inventory with a trusted colleague. What do you already do that you will keep doing? What will you do next to improve in the area of the learning environment, communication with parents, or adjusting mathematics instructional routines and tasks to show the value you place on multilingualism?

CONCLUSION

Centering instruction on the strengths and knowledge of students requires us to open up instructional spaces for students to engage in deep and rich mathematics, while providing the right supports. At the same time, too much scaffolding can take away from the work. Culturally responsive mathematics teachers first start from strategies that center children's brilliance and multiple forms of knowledge, in particular the vast linguistic as well as cultural knowledge bases children have. We then scaffold children carefully, through social and analytic scaffolds, so that they are able to engage in the mathematics while also learning mathematics in deep, connected, and meaningful ways. We introduced a variety of strategies for determining and maintaining the cognitive demand of tasks, while also broadening what counts as valuable mathematical knowledge. We introduced explicit scaffolding techniques, as well as a word of caution about presuming children need a lot of scaffolding up front (e.g., hold some of those in reserve until necessary.) And finally, we discussed the negotiation of more than one language in the mathematics classroom and how to affirm multilingualism and position all students as central to the learning. In the next chapter, we dive deeper into power dynamics that emerge as you make your mathematics classroom a space where high expectations with high supports and attention to positioning of students takes center stage.

DISCUSSION QUESTIONS

- How were rigor and support already part of your classroom instruction?

- What ideas in this chapter resonated/stretched your thinking about mathematics instruction?

- What activity ideas from this chapter will you initiate as part of your professional learning journey in CRMT?

CHAPTER 5

....

POWER AND PARTICIPATION

Knowledges & Identities	Rigor & Support	Power & Participation
Centering Cultural and Community Funds of Knowledge	Sustaining High Cognitive Demand	Distributing Intellectual Authority
(Re)Humanizing Mathematics	Scaffolding Up	Disrupting Status and Power
Honoring Student Thinking and Ideas	Affirming Multilingualism	Analyzing and Taking Action

Let's check in on Alicia one more time:

Alicia, the third-grade teacher in the Mission district of San Francisco, is at her desk reflecting on the math lesson dynamics that occurred in class today. Some groups worked collaboratively while other groups needed more practice listening to each other's ideas. There was one student, Olivia, who was quiet most of the class. Alicia noticed that during the explore part of the lesson, Olivia was creating arrays to help solve computational and story problems involving one-by-three-digit multiplication. However, Olivia was not sharing her math ideas with her group mates. "Olivia has yet to shine," thought Alicia. Alicia took out her class seating chart and started moving names around. She noted, "I need to think about redesigning our group work. We are starting our community action projects next week. This would be a good time to make some adjustments."

INTRODUCING THE STRAND OF POWER AND PARTICIPATION

Alicia's challenges with group dynamics are a common dilemma many teachers face when trying to broaden equitable participation in their classrooms. Power and positionality are always in play in mathematics classrooms. Consider the following questions: Whose voice is heard during math class (power)? Who is positioned to be powerful, and why (status)? And what power does mathematics have as a tool to help us analyze and make positive change in our world (action)? The focus on power is intentional, as mathematics classrooms, like all classrooms, are not immune to social forces that affect how students experience, interact, and learn mathematics with and from each other (Featherstone et al., 2011; Joseph, 2022; Martin et al., 2010). Since power dynamics exist in every classroom, how do these dynamics support positive and productive interactions, minimize status issues and disrupt stereotypes, and invite children to take action to address inequities and injustices with mathematics? In this chapter, we focus on the third strand of the Culturally Responsive Mathematics Teaching Tool (CRMT2): *Power and Participation*. You will learn about the various ways power and participation impact how we teach and learn mathematics.

BIG IDEAS IN THIS CHAPTER

- When intellectual authority is distributed, more voices, ideas, and actions can be leveraged for individual and collective mathematics learning.

- Deliberate compassionate focus on student strengths can disrupt entrenched stereotypes and power dynamics that block meaningful mathematical learning.

- Children can be change-agents with mathematics—making a positive impact on themselves, their friends, families, communities, and the world around them.

We will introduce the three dimensions that make up the strand: Power and Participation:

7. *Distributing Intellectual Authority*

8. *Disrupting Status and Power*

9. *Analyzing and Taking Action*

As in previous chapters, each dimension's rubric is presented, with a description of the dimension, ideas for how to attend to the dimensions in mathematics instruction, and how to interpret the rubric. In addition, every dimension is accompanied by a short teaching story and an activity for your own professional learning. Given entrenched inequities that persist inside and outside of the mathematics classroom, we invite you to build on your instructional strengths and stretch your practice to address those inequities/injustices. You can create a mathematical learning community that is powered by heartwork and compassion.

DIMENSION 7: DISTRIBUTING INTELLECTUAL AUTHORITY

ESSENTIAL QUESTION

How does my lesson distribute mathematics authority and make space for multiple forms of knowledge and communication?

This dimension is about relinquishing control over knowledge in the classroom, opening space for students to feel ownership and agency in their learning, and expanding what counts as valuable mathematical knowledge. When intellectual authority is grounded solely with the teacher or the text, learning is one-dimensional and limited. Paulo Freire (1970/1993) referred to this limited approach to teaching as the banking education model. In this model, students are empty vessels where tightly controlled knowledge authored by the teacher is to be deposited into students rather than co-constructed with students. We are all likely to be familiar with the idea of the teacher-centered classroom, where the teacher is the one who elicits and evaluates ideas, in what has come to be known as the IRE sequence (Initiation, Response, Evaluation: teacher initiates, students respond, teacher evaluates). The teacher controls the knowledge and learning. As research on classrooms has shown, when teachers delegate intellectual authority to students, there is deeper engagement in learning (Dunleavy, 2015; Webel, 2010). Langer-Osuna (2018) argues that the perception of who has authority in the classroom also largely impacts learning because it influences who students listen to and whose ideas they think have merit. Moving from a teacher-centered classroom to a student-centered classroom is one way to increase students' mathematics authority. However, attention to the *distribution* of mathematics authority over time is also important. That is, over time, authority must be widely distributed so that all students see themselves and each other as authorities. This means that students have frequent opportunities to communicate what they know in multiple modalities, including languages; at the same time, teachers attend to who holds status or power in the classroom. For example, teachers might ask themselves if there are particular students who everyone sees as "smart" or "right"? If so, teachers need strategies to broaden that status so that more students see each other as

powerful math thinkers and doers. Participation structures that are useful for *Scaffolding Up*, like "think-pair-share" along with collaborative group work, are strategies that help maximize student contributions. Those contributions are heard and valued by classmates and the teacher. Through widely distributed student-centered discourse, students engage more deeply with mathematics and with each other.

Instructional Approaches to Distributing Intellectual Authority

Distributing Intellectual Authority includes positioning students as mathematical authorities in the classroom, using participation structures that expand student mathematical communication opportunities, and embracing the role of teacher as facilitator and co-constructor of knowledge in the classroom. Some approaches include the following:

1) **Position students as holders of math knowledge.** We may all be familiar with students being *contextual* authorities. That is, they are experts on their own lives and situations they have encountered. But we may be less familiar with recognizing them as *intellectual* authorities in light of their backgrounds and experiences. With CRMT, intellectual authority is distributed between teacher and students and among students. The teacher and text may facilitate the development of mathematical knowledge, but neither own nor control that knowledge. Knowledge is generative with multiple perspectives and strategies offered to solve mathematical problems. Students can only show what they are capable of if the modes of engagement in the classroom facilitate opportunities to be mathematics authorities.

2) **Expand use of collaborative participation structures.** Familiar structures such as "turn and talk," "think-pair-share," "collaborative group work," and "gallery walks" provide opportunities for students to voice their ideas in multiple ways, listen to each other's ideas, share insights about their work, and stretch their curiosity by asking each other questions. These types of structures create collaborative intellectual spaces that promote learning, connection, and respect. However, sometimes well-meaning strategies such as "equity sticks" and "randomizers" may inadvertently create inequities. For example, what if you happened to have a class that has a majority of children from a specific demographic group? This can happen in "gifted" education settings where white children are overrepresented or in

English-dominant classrooms with few multilingual students. If you use an equity sticks approach where you randomly pull a popsicle stick with a student name from a can, then you may end up over time positioning students with privilege as the intellectual authority in the classroom and keeping the voices and ideas of nondominant students at the margin. It takes deliberate attention and pedagogical action to create equitable spaces that distribute intellectual authority broadly. It's crucial to reflect on your methods to make sure that multiple student voices and distributed intellectual authority are centered, creating a space where students feel empowered.

3) **Embrace the teacher-as-facilitator role.** *Distributing Intellectual Authority* means a different role for teachers—one that focuses on facilitation and co-construction of knowledge. This means that it is important for teachers to examine and understand their role in the math learning space. Are you taking up more space than students? Do you need to step back and encourage productive struggle? Do you invite exploration and curiosity? Can you ask a question rather than validate a solution? In addition, a facilitator role also means attending to how students are taking up space in the classroom. Being aware that some groups or individuals are taking on a more authoritative role calls for the teacher to disrupt and redistribute that authority. We will address this point as part of Dimension 8.

Understanding the Rubric for Dimension 7

The rubric is designed to help you understand if the lesson provides ample opportunity for intellectual authority to be widely distributed among students, centering student voices and multiple modalities of communication, or if this is a growth area because knowledge construction is teacher-controlled restricting student participation. The essential question for *Distributing Intellectual Authority* is *How does my lesson distribute mathematics authority and make space for multiple forms of knowledge and communication?* Figure 5.1 shows the rubric for Dimension 7.

At rubric Level 1, student participation is limited, such as in an initiation-response-evaluation (IRE) pattern. Classroom interactions are controlled

FIGURE 5.1 Rubric for Dimension 7: Distributing Intellectual Authority

Fragile/Margin				Strong/Centered
1	**2**	**3**	**4**	**5**
The authority of math knowledge exclusively resides with the teacher (e.g., teacher tightly controls talk in the classroom; teacher decides what answer is correct; IRE patterns may be evident in classroom discourse). Student participation is severely limited (e.g., limited to one-word answers, short choral responses, repetition of teacher, etc.).	The authority of mathematics knowledge is infrequently shared and primarily resides with the teacher and a few students. Student participation is limited (e.g., limited to one-word answers, short choral responses, repetition of teacher, etc.).	The authority of math knowledge between teacher and students is sporadically shared and resides with teacher and some students. Some students participate in math activities in substantive ways, periodically sharing reasoning and different strategies, and understanding the strategies of others.	The authority of math knowledge is equally shared among teacher and many students. Most students participate in mathematical activity in substantive ways and frequently communicate mathematical ideas in at least two modalities (e.g., listening, writing, drawing, speaking, gestures, etc.).	The authority of math knowledge is widely shared among teacher and most students, and *students* hold most of the math authority. All students participate in mathematical activities in substantive ways and communicate mathematical ideas through multiple modalities (e.g., listening, writing, drawing, speaking, gestures, etc.).

by the teacher. And the student responses are tightly structured with limited room for explanation or analysis. At Level 2, intellectual authority remains primarily with the teacher but might include a few students. The teacher tends to call on the same small group of individual students for the "correct" answer, leaving little room for other voices to make mathematical contributions. At Level 3, we begin to see a shift in the intellectual authority from the teacher to more students in sustained ways. This shift is sporadic, perhaps happening during the launch of a lesson, but there is some attempt to include more students in making mathematical contributions and an attempt to expand what counts as valid and valuable mathematical contributions. At Level 4, clear structures are put in place to distribute the intellectual authority equitably among the teacher and students. This means that the teacher structures ways in which students pose questions, offer ideas, analyze strategies, and communicate mathematical ideas in multiple ways, not just one way. The intellectual authority is more distributed with the teacher taking a facilitator role, rather than an authoritarian role. At Level 5, there is classroom-wide access to multiple modalities for engaging in, expressing, and analyzing mathematical ideas. This could include students using models to show their thinking; gestures and manipulatives to think through ideas; writing about mathematics, such as agreeing or disagreeing with a statement and explaining why; or providing opportunities to respectfully critique the ideas of others by offering compliments and questions on each other's work. The teacher is clearly a facilitator, finding ways to maximize engagement opportunities through various approaches to distributing authority.

Teaching Story: Broadening Mathematical Authority

Angelica is a fourth-grade student and a recent immigrant from Honduras. She solves an equal-sharing problem of 11 brownies shared among 3 people, so that each person gets the same amount, and none is left over. Her teacher, Ms. Nguyen, has introduced equal sharing problems to better understand how children think about partitioning and foundational fraction concepts. Angelica's solution is that each person gets 3 brownies, and the 2 leftover brownies should be saved for later. In the context of equal-sharing problems with fractions in the mathematics classroom, we might say the last two brownies should be cut up and distributed equally among the 3 people. But rather than dismissing Angelica's thinking as a misunderstanding of the fair sharing situation (diminishing her intellectual authority), Ms. Nguyen decided to ask the rest of the class to consider why Angelica's answer makes sense and under what conditions. This move invites other students to consider how their own thinking compares with someone else's and how that other way of thinking is also mathematically sound. For example, Angelica's answer makes sense if we are looking for a whole number solution and we are okay with saving brownies for later. Then Ms. Nguyen asks another student who cut up the last two brownies to share. She then invites Angelica to compare this solution to her own—how are they alike, and how are they different? This is an opportunity for Angelica to consider other ideas in relation to her own without suggesting that her solution is wrong or incomplete. She steps firmly into the role of mathematical authority for a moment. Angelica notices that her own solution has no cut up brownies, but the solution of her classmate has two brownies that were cut up.

Next Ms. Nguyen asks Angelica to consider if there are situations in which it makes sense to cut up the last 2 brownies and how her knowledge of fractions might help with this situation. Angelica considers that it might be okay to cut up the brownies if you need to use them all up. Other students also agree that when you have to use them all up, it's best to cut up what is being shared. The teacher could choose to keep the conversation going among the class as they collectively expand their thinking about what mathematical reasoning applies to this situation and to other related situations. Here the teacher is able to act as a facilitator to aid in distributing mathematics authority, which stems from sense-making and experience with both division and perhaps from Angelica's life.

It is important for culturally responsive mathematics teachers to consider not just who has mathematics authority but also what counts as mathematics authority in the specific context of their classrooms. Culturally Responsive Mathematics Teaching (CRMT) provides frequent and sustained opportunities for children to dialogue with each other using their mathematics authority as a way to deepen their own mathematical understandings when there are different strategies and ideas offered. By broadening the distribution of mathematical authority, students feel empowered to make mathematical contributions and build on each other's ideas, enriching the mathematical experience for everyone.

"By broadening the distribution of mathematical authority, students feel empowered to make mathematical contributions and build on each other's ideas, enriching the mathematical experience for everyone."

ACTIVITY: Mapping Mathematical Authority

Critically reflecting on the participation structures in our classrooms gives us insight into how distributed intellectual authority might be. When we make a mathematics authority map, we get a sense of who is seen as holding mathematical authority and who does or does not see themselves as a producer of mathematical knowledge.

Making an intellectual authority map can be done by making a *sociogram* of how children see each other in relation to mathematics. Sociograms have been used for decades to help teachers and researchers make sense of peer-to-peer relationships. We have adapted the sociogram activity from Shagoury and Power (2012):

1) **Conduct interviews with students.** If your classroom has older children, you might do this as a survey, but the classic method is to conduct very brief interviews. Ask each child, "Who would your first pick for a playdate be in this class? Who would your first pick be for working together on a math project? Who would your first pick be to read together?" You may also want to ask other questions of interest, such as "Who is really good at math in this class? Tell me your top three. How do you know they are good at math?" Write down all the responses from each child. You might also find it interesting to record them on a grid such as the one below.

2) **Create a math authority map.** Once you finish the interviews, you can start to total up who has the most votes for each category. You may decide to focus on one variable at a time, such as who selected whom to be a partner for math work or who each person indicated is their number one student. Write their names, and then use arrows and write the names of students who selected them as you start to map out your classroom relationships. This part might get messy, so you may want to do this on big poster paper with a variety of colors or marker patterns to help you track relationships.

3) **Analyze the map.** Who do students see as authorities in your classroom? Which students are isolated (meaning they don't have connections to others)? Are there any cliques, and if so, how can you tell?

4) **Strategize for change.** Often sociograms reveal that certain students are seen as authorities more than others. We might notice cliques of students who like to work only with each other, or we might notice students who are outcasts, not seen as authorities. Also, when we start to explore classroom data, we start to ask new questions as well: "What is influencing how students see each other? Is it the groups I have them in? Is gender/race/language playing a role? Are friendships?" With new information from the authority mapping activity, you can look at your classroom in a new way and decide what techniques you will use to redistribute authority.

Here is an example of a teacher mapping mathematical authority:

In this example, a second-third-grade combo teacher, Mx. Aguilar, decided to see how relationships among the twelve third graders were developing. They have interviewed seven students so far, and Arturo will be interviewed next. On the left there is a list of the students interviewed (Figure 5.2). Across the top are the names of all students in the class. P is used for playdate, R for reading buddy, and M for math preference. Numbers 1, 2, 3, in this case, represent how children responded to "who are the top three math students, and how do you know?" In this snapshot of some of Mx. Aguilar's data, what patterns do you observe?

FIGURE 5.2 Mx. Aguilar's Classroom Data

I ▼	Erica	Shawn	Tardell	Lia	Nisya	Manifest	Paul	Arturo
Erica	--				P/R			M 1
Shawn		--	P					M 1
Tardell		M	--		R			
Lia				--		P		
Nisya	P/R/M				--			
Manifest				P/M		--		1
Paul							--	M/ 1
Arturo								--

Mx. Aguilar's classroom data from student interviews about who their top choices are for different activities.

Next, Mx. Aguilar takes a moment to map out just the *M*s, which are the responses to *Who would be your first pick to do math with?* This gives them a different way to look at what the students are expressing about who they prefer to do math with. As you look at the map (Figure 5.3), what connections among students do you notice? Who is emerging as a person with authority? Who is not, yet?

Like Mx. Aguilar, you may already have hunches about who in your classroom is positioned as a mathematics authority and who is not. The map helped them see more clearly what connections exist between students, who is positioned as an authority, and what groups of students might be isolated from others. Mx. Aguilar continued the interviews and expanded the data set and authority map. Reflecting on the connections between children, they made a plan to more explicitly distribute the intellectual authority, which includes asking students their ideas again (through another interview or a survey) later in the year.

This kind of analysis can provide insight about distribution of math authority as well as power and status in the classroom.

FIGURE 5.3 Mx. Aguilar's Map

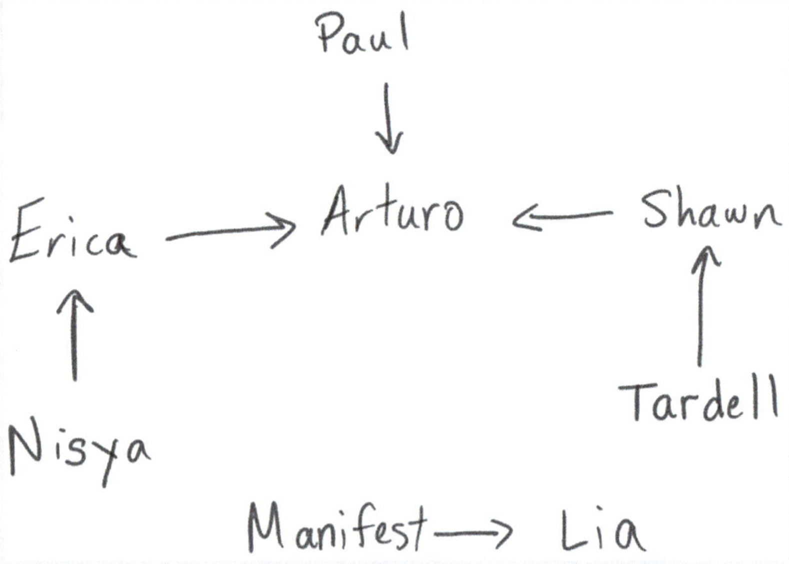

The map Mx. Aguilar has created so far from student data.

DIMENSION 8: DISRUPTING STATUS AND POWER

ESSENTIAL QUESTION

How does my lesson disrupt status differences, entrenched stereotypes, and inequitable power relationships present in all mathematics classrooms?

In the work of Cohen and Lotan (1995/2014), status issues are defined as inequities that arise in the classroom and impede some students' ability to learn because certain students hold more power while other students hold less. Status can be attributed based on a number of characteristics. The primary categories we may be familiar with are social status and academic status. Social status refers to a hierarchy of value or worthiness based on looks, popularity, and perceived talent of a person in a classroom. Academic status is a hierarchy based on valued forms of performance in an academic content area, such as mathematics. While social status and academic status overlap, it is generally agreed that they hold different currency. Academic status is often linked to ideas of smartness. As students figure out who they believe is smart in their mathematics classes, they also may attribute more social status to those students.

Some status issues that play out locally in the classroom can be disrupted through carefully planned instruction. Complex instruction leverages multiple and diverse strengths of children through collaborative learning (Featherstone et al., 2011; Horn, 2012). Using group-worthy tasks that are mathematically rich, complex, and demand collective engagement and then publicly assigning competence to various mathematical contributions are two particular strategies that can help students to see each other as mathematically competent. Growth mindset is nurtured in these classroom settings with strong learning outcomes and positive postsecondary education trajectories (Boaler, 2015; Dweck, 2006).

However, a critical look at status reminds us to also account for the differences between local meanings about status characteristics and bigger discourses, such as systemic racism, sexism, and classism, related to mathematics achievement that also play out locally in all math classrooms. Racialized discourses about who can be good at mathematics are not necessarily dismantled by a growth mindset. Rather, challenging the power of social hierarchies in mathematics, so that all students can see themselves and each other as mathematical people, requires explicit attention to race, class, and gender in the mathematics classroom (Zavala & Hand, 2019). For example, when teachers engage in test-score-driven "data dives," there is sometimes a hyperfocus on standardized test scores and "gaps." The results of these assessments (i.e., kindergarten readiness tests, math placement tests, state-standardized tests, I.Q tests) lead to deficit labeling of children and constructing status in the classroom (Gutierrez, 2008).

As soon as students are labeled with terms such as "low," "low-performing," "mid-range," "average," "above average," "high," "bubble," "gifted," or "highly capable," structures are put in place that reinforce those labels. Education for those groupings is differentiated as well as racialized, gendered, and classed. Ability grouping and tracking reinforce these labels leading to different education experiences for students called *academic apartheid* (Berry et al., 2014; Wells, 2018). For example, for students placed in interventions designed to remediate, mathematics curriculum often consists of low-cognitive-demand tasks and fact-fluency practices that are tightly controlled and monitored. Few opportunities for analysis and problem-solving are provided by teachers because there is an underlying belief that students exhibiting below-grade-level competencies must "master their basic skills" before they can engage in complex problem-solving. In contrast, students placed in classes designated as "enrichment," "advanced," "gifted education," "International Baccalaureate (IB)," or "college prep" often engage with high cognitive demand, inquiry-based, and nonroutine tasks with collaboration and creativity explicitly encouraged. On-ramps to these tracked course pathways are well-designed. Off-ramps are few by design. While a focus on growth mindset is admirable, it is incompatible with tracking because tracking is the structural equivalent of a fixed mindset (Boaler, 2015; Dweck, 2006).

Our current math education system is set up to rank, sort, and place children using standardized assessments that, unfortunately, trace back to the eugenics movement of the early 20th century—a research-based educational hierarchical system by race, class, and ability that placed able-bodied white people and the wealthy class at the top and Black and Indigenous people, poor/working class, and people with disabilities at the bottom (Berry et al., 2014). Children and their families are not immune from these historically entrenched status messages. Children can often tell you which classmate is "smart in math." They can articulate their own identity in relation to this manufactured smartness. Young children can also depict who is good at mathematics through drawings. Child-generated pictures of mathematicians and scientists can speak volumes about how children see themselves in relation to mathematics (Tucker-Raymond et al., 2007).

In addition, the hyper focus on gaps often centers on why certain student groups are not doing well in mathematics, without acknowledging what resources might already be in place that advantage students who do well on traditional standardized measures, such as access to technology and private tutors, shadow education and test-prep businesses (e.g., Kumon, Sylvan), and summer and after-school enrichment activities (robotics club; art/music classes, choir, theater). By minimizing these advantages, there is a misguided focus on student intellectual abilities as measured by standardized tests as if everything is equal when it is not, by design. Today's prevalence of these rigid structures perpetuates race, gender, and class disparities and stereotypes about mathematical advancement. As a result, children of color, children impacted by poverty, children with disabilities, and girls are often shunted out of STEM (science, technology, engineering, and math) education by (1) denying access to mathematical enrichment; (2) through isolation—being the only or one of the few and dealing with constant micro and macro aggressions about their competence; or (3) being rendered invisible—especially for Black girls (Berry et al., 2014; Gholson, 2016; Joseph, 2022; McGee, 2020). Therefore, as culturally responsive mathematics teachers, we have to also explicitly disrupt entrenched patterns of marginalization and stereotypes in the mathematics classroom, with attention to the power these patterns hold inside and outside of our own classrooms.

Instructional Approaches to Disrupting Status and Power

CRMT is incompatible with the structural equivalent of a fixed mindset—tracking. And since tracking reflects systemic forms of racism, sexism, classism, and ableism, the CRMT2 is a helpful tool to dismantle systemic oppression that affects how children experience mathematics learning in and out of school. Minimizing status differences in the classroom starts with acknowledging that status differences are structured in the education system and taking steps to change. As teachers, we can address this situation by taking an asset-based approach to our instruction. Below are three strategies to help make change:

1) **Humanize assessment.** Current uses of assessment tools contribute to the dehumanizing labels assigned to students because they mainly point out what kids can't do. We need assessment tools to be formative, holistic, and balanced. This kind of assessment approach helps us identify strengths and areas of growth for each child we work with because each child has them—strengths and growth areas. Even if you work in a district that prioritizes standardized test performance, by expanding the ways you assess, you can provide a more accurate and humanizing portrait of what your students have learned and what they are still learning to do. A key here is being able to identify strengths (what a child can do) and then use those strengths to bridge their developing competencies.

2) **Assign competence.** One way to do this is by focusing on a student who has yet to shine in your class. Take a few days and deliberately look for strengths related to mathematics and to other areas such as leadership, language use, listening, helping, problem-solving, and artistry. Note when this student is exhibiting strengths, and publicly assign competence to this child's idea or action. For example, here are some teacher moves (www.eqstemm.org) that might help you do that:

   - Hannah just shared a really interesting wonder. Let's see if we can build on their idea to decide on our problem.

   - Ji-Yeong, you helped us understand why this situation is important. What questions should we ask?

   - I see you created a picture of the situation. Can you tell us what this part of the picture represents?

   - This is a really unique way to represent a math model. Let's all learn from these ideas.

   - I really like how you solved this problem. Can you share your strategy with the class? We learn from your ideas.

   - Abud's solution was different. Let's listen and think about the different strategies they used to solve the problem.

By assigning competence to students who may suffer from societal stereotypes related to race, gender, class, or ability, you increase the likelihood of those students gaining confidence in themselves as doers of mathematics (i.e., math identity). You also model for children ways to acknowledge and value contributions of ideas from a variety of sources, positioning all to learn from one another (***Distributing Intellectual Authority***) rather than only a focused privileged few.

3) **Co-construct group norms.** As first described in Chapter 4, fostering collaborative learning requires specialized attention to norm building so students see, experience, and value learning from each other's ideas. It is especially helpful to co-construct group norms to help teams be productive and respectful. Norms are meant to be revisited and revised throughout the school year. There are various examples of norms that help group dynamics. The following is an example of norms that were co-constructed in a fourth-grade class. These norms were revisited before the start of math lessons, posted on the wall for reference, and often put on math activity sheets to remind students of the expectations.

- Listen to understand.
- Everyone participates.
- Everyone stays together.
- Everyone agrees upon a question before asking the teacher.

An asset-based approach to mathematics instruction is at the heart of CRMT. This can be done in a variety of ways and settings. By focusing on strengths and collaboration, power dynamics can be changed from unproductive to productive, helping all to see the multiple strengths children have and are developing in their math learning space.

Understanding the Rubric for Dimension 8

This dimension draws teachers' attention to the use of specific strategies to both engage all students and attend in particular to challenging stereotypes of who can engage in rigorous mathematics. The evidence centers around strategies that aid students to engage with each other as equals and also explicitly address negative stereotypes through building understanding, compassionate talk, and ultimately cultivating cultural humility and empathy across the mathematics classroom. The essential question for ***Disrupting Status and Power*** is ***How does my lesson disrupt status differences, entrenched stereotypes, and inequitable power relationships present in all mathematics classrooms?*** Figure 5.4 shows the rubric for Dimension 8.

At Level 1, the teacher is not working to disrupt status issues, especially those that may stem from particular known stereotypes in mathematics. Across Levels 2 to 5, explicit use of strategies to disrupt status issues grow from one or a few to many sustained strategies at work. In parallel, teachers pay more attention to involving students from a variety of backgrounds and experiences. In particular, we encourage teachers to consider how their

FIGURE 5.4 Rubric for Dimension 8: Disrupting Status and Power

Fragile/ Margin → Strong/ Centered

1	2	3	4	5
No strategies to minimize status issues are evident.	At least one strategy to minimize status differences is evident but superficial and does not challenge stereotypes or other power dynamics.	Some strategies to minimize status differences among students (and specific subgroups) in the lesson are evident and have some effect.	Some strategies to minimize status differences among students (and specific subgroups) are evident and have some effect.	Multiple strategies to minimize status differences among students (and specific subgroups) are implemented effectively throughout the lesson.
Student involvement is structured to privilege a dominant subgroup (in terms of race, class, gender, language, (dis)ability, and other socially constructed identities).	Student involvement is structured to privilege a dominant subgroup (in terms of race, class, gender, language, (dis)ability, and other markers of status), with limited involvement from nondominant students.	Strategies may have a momentary impact on some subgroup but may not necessarily address a persistent status issue related to race, gender, (dis)ability, language, and other markers of privilege.	Teacher uses one or more strategies that • maximize student mathematical, cultural, and linguistic strengths, • explicitly address stereotypes, and • structure compassionate and inclusive talk (i.e., building each other up, not tearing down)	Teacher and students both work to minimize status issues through strategies that • maximize student mathematical, cultural, and linguistic strengths, • explicitly address stereotypes, • structure compassionate and inclusive talk (e.g., building each other up, not tearing down)
		Student involvement is structured to support particular subgroups, which may include some but not all nondominant groups.	Student involvement is structured to support most nondominant subgroups.	Student involvement is structured to support multiple or all subgroups, with particular attention to historically marginalized and segregated students.

implicit bias and dominant narratives of who can be good at mathematics might impact their abilities to engage students from various racial, linguistic, (dis)abilitied, gendered, classed, and other social identity subgroups of students. Finally, this dimension is also about explicitly disrupting damaging stereotypes stemming from racism and other systems of oppression *during* mathematics instruction. Across the rubric, we can see a progression of teaching from a color-evasive stance toward teaching with awareness of racialized discourses and attending to the complex ways race and racism inform learners' identities in mathematics. Similarly, attention moves across the rubric from privileging only dominant groups' ways of knowing and participation, toward positioning knowledges and participation of nondominant students as equally valuable and assets to learning.

Teaching Story: Disrupting Status and Power

Ms. C observes the third graders bustling with excitement as they work in groups to figure out how much water would be enough for their class for one day if their school water was contaminated. One specific group caught her eye. A Black boy named Thomas comes in and out of the group's dynamic. Thomas had recently joined the class a few months ago. He was from the city but was new to the neighborhood. He was still in the process of making friends and building trust with her and his classmates. She watches as Thomas approaches the group, shares an idea, then steps back from the group, and takes a lap around the room. As Thomas backs away, his teammate states, "Oh, I get it. Thanks, Thomas." Ms. C notes that this was the first time Thomas fully engaged in a math activity. This was a success to celebrate. He rarely interacted when it was a textbook lesson. When he returned from his lap, Ms. C approached Thomas and asked him to help her understand an equation on the group's poster. Thomas quietly explained that the group was keeping track of the number of people in the class and how much water they drank per day to get the total number of cups of water the class needed. Ms. C smiled and thanked him for the explanation. Then she said, "Thomas, that was an excellent explanation. When your group presents, would you mind sharing that explanation to the class? We can learn from you." He looked at her and shrugged, "I guess so, sure." Ms. C smiled and began to transition the class to group presentations.

It might be easy to attend to what Thomas was not doing such as consistently working with his group. It also might be easy to miss or dismiss his contributions if solely focused on behavior that might be interpreted as off-task such as taking laps around the class. In fact, Black children are more apt to be disproportionately disciplined for such behaviors than white children (Children's Equity Project, 2022; Crenshaw et al., 2015; Welsh & Little, 2018). But by focusing on strengths, Ms. C observed that Thomas was making valuable contributions to the group, and he could explain their emerging ideas. The group also acknowledged Thomas's contributions. His ideas were offered verbally and helped the group make progress. Other group mates were building on these ideas and constructing the poster for the group presentation. Different group strengths were being leveraged. Ms. C knew that Thomas sometimes needed to remove himself from class activities by taking a lap or sitting in the reading space to tap down anxiety. Ms. C carefully selected group mates who showed strengths in patience and listening skills to work with Thomas. Looking at strengths enabled Ms. C to assess progress for individual and collective learning. Assigning competence positioned Thomas as an intellectual resource for the class.

PAUSE AND REFLECT

- Mathematics is fraught with stereotypes about who can and can't learn. How do you challenge stereotypes related to mathematics? In what ways do you position students who have yet to shine as mathematical resources for the class?

ACTIVITY: Telling New Stories

This dimension focuses on *Disrupting Status and Power* in the classroom, including entrenched stereotypes that affect how we are positioned and position others in the world. Part of this work is claiming our own stories and histories and understanding how some of the stories we tell about students are based in stereotypes or oversimplified stories of how the world works. Rather than subscribe to the same old narratives about who is supposed to be a mathematician, we need to tell the new stories, the *counter*stories that help challenge the old. Here are two story creation activities that might help you along in this journey.

1) **CREATE A MATH ORIGIN STORY.** It is helpful to start with your own math identity and how you experienced mathematics growing up. Creating a math origin story helps with identifying ways power, privilege, and oppression affected your math learning experience. Here are some prompts for you to create your own origin story. We invite you to share this story with your colleagues and perhaps your students as well.

MATH ORIGIN STORY

Part 1: Take some time to write/record your responses to the following prompts:

a. **Set the Stage**

- What were three critical moments in your journey to becoming a mathematics teacher?

- Who were two people who encouraged or discouraged you in this journey?

- What is one example in which you felt joy in learning mathematics?

(Continued)

b. **Digging Deeper**

- What roles did social identities (race, gender identity, age, dis/abilities, ethnicity, sexual orientation, class, religion, language, etc.) play in your origin story?

- Were you ever tracked into specific math pathways or groups? Was there status attached to you or the pathway? How did that make you feel?

- Were the people in your math classes similar/different from you?

- How was mathematics experienced in your family?

c. **Connect to Classroom Practice**

- What aspects of your own history with mathematics do you think have an impact on your views about teaching mathematics (i.e., your mathematics vision, your why)?

- What kind of math identity do you want your students to develop in your classroom? How might your actions impact your students' math origin stories?

Part 2: Tell your origin story with a representation you can share with a partner. This story can be represented with multiple modalities and is up to you: drawing, spoken word, writing, poetry, or so on.

2) **CREATE STUDENT STRENGTH STORIES.** There is so much attention to gaps and what students do not know that contributes to entrenched racist stereotypes about who can and who can't learn mathematics (Gutiérrez, 2008). We invite you to create counter narratives that focus on a student's strengths. Pick a child who has yet to shine in your class and is different from you based on social identities. Take time to get to know this student and his or her family more deeply. Observe this student while in class as well as social settings like lunch and recess. Start to identify strengths that are mathematical and in other areas such as leadership, friendship, artistic expression, and so on. Create an alternative narrative of this child that highlights strengths. Share this information with the child and his or her family.

DIMENSION 9: ANALYZING AND TAKING ACTION

ESSENTIAL QUESTION

How does my lesson support student use of mathematics to analyze, critique, and address power relationships and injustice in their lives (economic, social, environmental, legal, political, patriarchal)?

The final dimension of the CRMT2 is focused on how students use mathematics to make sense of the world around them and take action as community change-agents. This dimension is about multiple layers of power. Specifically, it is about students getting opportunities to analyze power relationships in the world and disrupting traditional notions that prioritize school math above family and community math. Using mathematics as an analytical tool to examine equity and justice can mean investigating situations of fairness, representation, access, civil rights, and bias. When students explore the ideas reflected in this dimension, they develop a sense of critical consciousness and critical civic empathy about the world around them and their place in it (Mirra, 2018). This means that not only is it important to understand and make sense of a situation, students must also understand how power, privilege, and positionality are involved; that personal experiences matter in the situation; and there are specific opportunities to take civic action to address equity and justice. CRMT teachers affirm that what matters to students takes precedence over the textbook generalities and that mathematics is best learned through an authentic context of posing and addressing meaningful problems (Aguirre et al, 2022; Tate et al., 2022; Turner et al., 2022).

There is a growing literature base of community-focused math tasks that emphasize analysis of power relations with mathematics. In addition to Gutstein and Peterson's groundbreaking book *Rethinking Mathematics* (2005, 2013), we recommend the PreK to grade 12 series *Mathematics Lessons to Explore, Understand, and Respond to Social Injustice* published by Corwin as a starting place for guiding principles, instructional practices, and specific math tasks that students can do across grade levels and inspire you to create your own tasks. In addition, the *TEEM Special Issue on Social Justice in Mathematics Education* (Aguirre & Civil, 2016) has articles with examples from classrooms on **Analyzing and Taking Action**, and the *TEEM Special Issue on Antiracism in Mathematics* (Zavala & Simic-Muller, 2022) also has examples of both **Analyzing and Taking Action** and attending to **Disrupting Status and Power** (see https://www.todos-math.org/teem).

We also know from research that when lessons are grounded in community/cultural funds of knowledge, there are more opportunities for teachers to address issues of power and social justice. In other words, local issues *are* justice issues (Aguirre et al., 2012; Turner et al., 2022; Zavala & Stoehr, 2019;). Thus, it makes sense that the dimensions **Cultural/Community Based Funds of Knowledge *(Strand 1, Dimension 1)*** and **Analyzing Power and Taking Action *(Strand 3, Dimension 9)*** are the dimensions that start and end our tool. By clearly anchoring ourselves in students' communities and funds of knowledge, we are likely to find opportunities to position students as change-agents for their communities. Below is a student-generated list of change-agent actions:

I AM/WE ARE CHANGE-AGENTS

★ Making friendship bracelets to welcome refugee families to our community

★ Helping families that are unhoused with a toy drive

★ Donating books to build a diverse library collection for our class and the apartment building next to school

★ Donating supplies for a local pet adoption event

★ Creating a community garden to help our community and local food banks

★ Beautifying our school with planting flowers

★ Writing letters to children in hospitals

★ Reducing cafeteria waste

★ Cleaning up trash around the school and local parks

★ Recycling plastic bottles into art projects and eco bricks

★ Upcycling plastic bags into jump ropes for our P.E. class or local community center

Instructional Approaches to Analyzing and Taking Action

To fully engage this dimension of CRMT, it helps to look at different ways current events are discussed in school. What happens if there is excitement about a new park or community center being built? What if there is a tragedy in the community—shooting, fire, accident? How do schools prepare students to be environmental stewards or be ready for natural disasters? How can we welcome newcomers to our community? How do we talk about elections, laws, and government? Children are both witnesses and actors in our world. They have questions, confusions, and ideas on how to make the world better. As CRMT teachers, it is important to not shy away from complexity and uncertainty. Supporting students to engage novel and complex situations builds critical consciousness (Ladson-Billings, 1995), which is a broader *sociopolitical consciousness* that allows students to "critique the cultural norms, values, mores and institutions that produce and maintain social inequities"—in other words, to engage the world and others critically (Freire, 1990; Ladson-Billings, 1995). Mathematizing those situations builds critical consciousness with mathematics. Below are three approaches to start this part of your CRMT journey:

1) **Deepen your critical consciousness.** To best support children becoming change-agents for their community, it is important for you to critically

reflect on your own power, privilege, and political perspectives related to the topic. It is most responsible to take time to deepen your critical consciousness prior to enacting a justice-focused task hastily. However, designing a justice-focused math lesson can be the catalyst for this work. Ask yourself questions such as who is best served by this situation? Who is not served? What is the underlying purpose, and how are power dynamics and social forces related to race, class, gender, language, and ability connected to this topic? There are several math focused resources to help you deepen your critical consciousness in relation to CRMT.

2) **Investigate current events.** Current events in the community, region, and world make powerful contexts for exploring mathematics in meaningful ways. You probably already engage in community-focused or service-learning projects such as community gardens, fundraisers, donation drives, pet care, recycling, food waste, water safety, and helping community members in need. Build on those, and find out about other needs in the community. Sometimes family liaisons, coaches, and nonteaching staff who work at the school can be good resources. In addition, community organizations, parks and recreation, community boards, local businesses, and faith centers can also be sources for ideas of current events impacting the community. But partnerships with local organizations must be undertaken with utmost cultural humility and respect. Learning for Justice's article on Anatomy of an Ally is a good starting point for considering how to enter spaces and practice allyship: https://www.learningforjustice.org/magazine/summer-2016/anatomy-of-an-ally

3) **Take action.** A hallmark component of this dimension of CRMT is taking action, which means students doing something as a result of their analysis to make a positive difference in the situation. This could be sharing their investigations via letter writing or presentation with people in power to help them make a decision (e.g., principal, school board, city board). It could be creating an infographic, mural or other artwork, or video for public display raising community awareness about the situation. If students are to be community change-agents, analysis is not enough. Students finding ways to make their ideas matter and be heard goes a long way in affirming their mathematical identities and making mathematics meaningful.

Understanding the Rubric for Dimension 9

This dimension draws teachers' attention to sociopolitical dimensions of current events and issues impacting communities locally and globally. Power dynamics are already at play. To help our students analyze and critique these power dynamics and become community change-agents, they must have the opportunity to investigate complex and uncertain situations with mathematics. Developing students' sociopolitical consciousness and taking action are centered. The essential question for **Analyzing and Taking Action** is *How does my lesson support student use of mathematics to analyze, critique, and address power relationships and injustice in their lives (economic, social, environmental, legal, political, patriarchal)?* Figure 5.5 shows the rubric for Dimension 9.

The rubric for this dimension moves from a teacher's own awareness and ability to incorporate analysis and attention to a matter of significance (Levels 1 and 2), to intentional support for noticing, analyzing, and taking action to address equity and justice (Levels 4 and 5). Particular to Level 5 is incorporating opportunities to build from the mathematics to take action as a change-agent for the community. Taking action could include making a presentation to a decision-making body such as a community-organizer group, school board, district/school leadership, or elected officials. It could also mean creating infographics about the situation and placing them in windows or bulletin boards educating the public about the situation.

Every teacher should expect that they do engage this dimension as part of their mathematics instruction *over time*, even in early childhood settings. This may also be an opportunity to see cross-curricular connections between social studies, science, and English language arts. We invite teachers to identify when they introduce their students to political, civil, and environmental justice leaders, particularly young people, who are fighting for equal rights (e.g., Malala Yousafzai); access to diverse children's books written by and for Black girls (e.g., Marley Dias); or saving our planet (e.g., Nanieezh Peter). Other community connections might be when teachers design class or schoolwide projects that benefit the community such as addressing food and housing insecurity, protecting our environment through recycling or building a community garden, or finding safe routes to school and back. How kindergarteners engage in issues of social justice in the mathematics classroom will no doubt look different from how fourth or fifth graders do. However, all children are capable of engaging with and in the mathematics of social issues and often eager to ask and answer questions that matter deeply to themselves and their communities.

Teaching Story: Analyzing and Taking Action

The Flint Water Crisis provided a learning opportunity for our nation's cities and schools to annually examine running water for contaminants. As a result, South Hill Elementary, a culturally and linguistically diverse school in Washington State, tested the school's water every spring. This time, lead was found in the water. Water fountains and sinks were declared immediately off limits until the pipes could be fixed. Families were sent letters that described the situation and how it would be addressed. Water dispensers with large (5 gallon) jugs of water were set up in the hallway for classrooms to share. Mrs. G, a fourth-grade teacher participating in a culturally responsive math modeling project, took this opportunity to use this real-world context for a

FIGURE 5.5 Rubric for Dimension 9: Analyzing and Taking Action

Fragile/
Margin

Strong/
Centered

1	2	3	4	5
No evidence of connection to critical knowledge (sociopolitical contexts, issues that concern students).	Opportunity to critically mathematize a situation went unacknowledged or unaddressed when present.	There is at least one instance of connecting mathematics to analyze a sociopolitical/cultural context, with the purpose of deepening understanding of how mathematics and the social issue connect.	There is at least one major activity in which students collectively engage in mathematical analysis within a sociopolitical/authentic or problem-posing context. Mathematical arguments are provided to solve the problems. Pathways to change/transform the situation are briefly addressed.	Deliberate and continuous use of mathematics as an analytical tool to understand an issue/context, formulate mathematically based arguments to address the issues, and provide substantive pathways to change/transform the issue.

math modeling task that would investigate how many jugs of water would be enough for the class for a day (Turner et al., 2019). She wanted her students to create an action plan that would help kids in other schools make sure they had enough water if something were to happen to their water supply.

To help build context, Mrs. G found scholastic news articles about the Flint water crisis. The students read the articles and discussed connections between what happened in Flint and their own situation. They looked at data describing the lead levels in children's blood before the crisis and during the public crises. On the next day, Mrs. G began the class with a mathematizing-the-world routine showing an image of two kids with a dolly pushing a large jug of water. The walls had several large jugs arranged on shelves, and there were water dispensers in the background. Mrs. G asked the students three questions: What do you notice? What do you wonder? And what questions can be answered with mathematics? Each time the children had an opportunity to talk in their table groups prior to responding. Mrs. G recorded their answers on an anchor chart. Children noticed the action of the kids pushing the large jug. One student observed that one child in the photo had "cool" gold sneakers. Students also noticed the jugs were arranged in "an array" on the shelf. One child observed that both kids in the picture were Black. There were many math-related wonderings children had as well. Children wondered how long the water would last. They wanted to know if there was enough water for the school. And, would they need more water dispensers? Mrs. G was excited about the problem-posing generated by the students as they directly connected to the posed task: Given that our water is not safe, how many large jugs of water are enough for our class for one day?

As the modeling lesson continued, Mrs. G asked students what they know about this situation, what they need to know, and what they could assume or decide. The goal here was to gather as much information from the students as possible. And if students identified something they needed to know, Mrs. G had information that would help them, such as how much water people of different ages need to drink per day as well as how many cups are in a one gallon. As children worked in small groups on the task, Mrs. G walked around the room listening actively. One child shared with their group that they had to use water bottles at home to cook their food and drink water. Another child said that they bring a water bottle to school and fill it up at least twice each day. As students created their models that included how many people would be drinking water and how much water they would drink per day, Mrs. G asked students to make sure they determined how many large jugs of water would be needed. As the activity wrapped up for the day, she reminded them that their plans would help other children in similar situations. They could be community change-agents that help others in times of need.

Mrs. G took an opportunity to create a math lesson connected to current events affecting the school community. She created a math activity that helped students better understand the importance of water to themselves and their community. She also wanted students to be change-agents who helped other students make a plan in case they had a water emergency.

ACTIVITY: Taking Action

Many of us are members of multiple communities through which we can enact change. Sometimes we might give donations. Sometimes we might go knock on doors for a cause. In the spirit of continued growth, this activity box is all about growing with(in) our communities. Here are a few things you can try to get you started in activating your activism:

1) **Brainstorm community-justice contexts.** Take a few minutes, and brainstorm community-based and service-learning projects you know about through your professional and personal networks. Then brainstorm curriculum topics that might lend themselves to taking action.

2) **Connect curriculum to community action.** Take some time to review units in other content areas. Are there connections to fairness, representation, access, and justice? For example, are there connections to nutrition, weather, pollution, population? Are there data tables, timelines, or graphs that can be analyzed with mathematics? Often, we may overlook cross-curricular connections that are mathematical in nature. Keep your eyes open for these opportunities.

3) **Adapt and teach a community justice math lesson.** There are justice-focused math activities already developed, such as the PreK to grade 12 series *Mathematics Lessons to Explore, Understand, and Respond to Social Injustice* published by Corwin. In addition, many American Indian/First Nations have websites and curricular resources that offer justice-focused lessons related to understanding the culture, language revitalization, and dark history of colonization, as well as protecting land, water, air, and earth. Some resources include Since Time Immemorial Curriculum (Washington State); https://www.k12.wa.us/student-success/resources-subject-area/time-immemorial-tribal-sovereignty-washington-state.

All of these resources are opportunities to engage students in critically analyzing our humanity in the past, present, and future. We invite you to take another look at your curriculum with fresh eyes to make connections to community-based and current events.

CONCLUSION

At the heart of the Power and Participation strand are ways to support student mathematical empowerment, agency, authority, and action. Doing that requires focused attention to how students are positioned as intellectual authorities in the classroom, keeping close tabs on the roles entrenched stereotypes and systemic oppression may play in classroom dynamics. Culturally responsive mathematics teachers embrace a facilitator and disrupter role by co-constructing classroom norms that leverage diverse ideas, strengths, and ways to communicate understanding rather than centering the math authority on the role of teacher, text, or a few "go-to" students. Math participation is centered rather than at the margins. Everyone is expected to participate and make valuable contributions to the collective mathematical learning. It is also expected that students will engage in critical analysis of important topics in their communities and use mathematics to take action and be community change-agents. The teaching cases provide illustrative examples for your professional learning. In addition, the activity boxes provide multiple opportunities for you to strengthen your understanding and practice your skill set with this strand of CRMT.

DISCUSSION QUESTIONS

- What roles have power and status played in your math classroom?

- What ideas in this chapter resonated/stretched your thinking about mathematics instruction?

- Which of the activities will you initiate as part of your professional learning journey in culturally responsive mathematics teaching?

PART 2

. . . .

CULTURALLY RESPONSIVE MATH TEACHING IN ELEMENTARY CLASSROOMS

In Part 2, we provide several examples of how elementary teachers used the Culturally Responsive Mathematics Teaching (CRMT) framework to plan, enact, and reflect on mathematics instruction. Our guest authors for each chapter were math teachers/coaches in a variety of elementary settings. Each chapter illustrates the flexibility of the CRMT framework through the voice and actions of each author. Chapter 6 shares the CRMT journey of an instructional coach, Holly Tate, and first-grade teacher, Olivia Canning, and their use of the CRMT2 framework to adapt a district-created math lesson to be more rich, rigorous, and relevant to students. Chapter 7 reflects the evolution of practice over four years through the eyes of an instructional coach, Holly Tate, and her teaching partner and fourth-grade teacher, Kaitlin Kaplewicz, to transform how students learn mathematics in more meaningful and culturally responsive ways. Chapter 8 focuses on culturally responsive teaching in bilingual settings in which guest author Melissa Adams Corral narrates her instructional choices and student strategies through a CRMT lens. Chapter 9, coauthored by Talya Kemper and Maria del Rosario Zavala, illustrates how CRMT can be enacted in two distinct special education spaces, a push-in model and a self-contained classroom.

Each chapter in Part 2 provides unique and flexible ways to use the CRMT2 framework to enhance and reflect on mathematics teaching practices. We understand that explicit examples of what CRMT looks like in various elementary settings can help teachers envision possibilities and take action. But we do not expect that the way each teacher here works with the CRMT2 in these moments is exactly how you will or is representative of the ways they themselves will continue to work with it as their own practices evolve. We invite you to learn from the stories in these chapters and the variety of ways teachers draw on the dimensions in the CRMT2 to plan, analyze, reflect, and repeat. Like the chapters in Part 1, there are frequent opportunities to pause and reflect on what you are reading and how these ideas might connect to your mathematics teaching.

CHAPTER 6

························

USING CRMT TO ADAPT A FIRST-GRADE DISTRICT-CREATED LESSON

By Holly Tate and Olivia Canning

In this chapter, two white mathematics educators working in an urban setting, mathematics coach Holly Tate and teacher Olivia Canning, detail their experiences in professional learning through the lens of the Culturally Responsive Mathematics Teaching (CRMT) framework. The job-embedded learning takes place through a coaching partnership in first grade. In this classroom, this coach and teacher worked in partnership to reconstruct a district-created lesson through careful planning using each of the CRMT2 dimensions. Let's hear from Holly and Olivia directly, as we explore the content and instructional decisions made and reflect on how those instructional decisions transformed the mathematics experiences for children.

TRANSFORMING A LESSON USING THE CRMT2

With eight years together as coach and teacher, we already had a long history of working as a partnership. Many of the children we worked alongside had, historically, not been given equal opportunities of access to rigorous mathematical thinking due to common assumptions of what they could or could not do as a result of neurodiversity, poverty status, and racial background. Throughout this chapter, we describe a new experience using the Culturally Responsive Mathematics Teaching Tool (CRMT2) framework as we focused on one mathematics lesson in Olivia's first-grade classroom. We were excited to provide students with some real-world math opportunities in a way that moved beyond story problems while also paying special attention to how we were becoming culturally responsive. We were looking forward to the opportunity to work closely with one another again, in an old coaching partnership, but through a new lens, to continue disrupting the status quo in a deeper way, through the use of the CRMT2 framework.

The demographic background of our first-grade classroom gives a snapshot into a diverse make-up: 40 percent of students identified as white, 20 percent as Hispanic, 16 percent as Asian, 15 percent as Black, and 8 percent as multiple races. Further, about 30 percent were multilingual learners, and 30 percent were receiving special education services. We began our first planning session with the district-created lesson plan for the first day of estimation and magnitude. Though we had worked together for years, we decided to tackle this specific lesson in a new way: through the lens of the CRMT2 framework. We wondered in what ways is the lesson, as is, culturally responsive? How might we need to adapt it to transform it into a more culturally responsive learning experience for students? Throughout this case, we touch on the strengths of the lesson and then expand on two big ideas that emerged through our planning. First, the need for lessons that focus on power and humanization. We unpack how we address those needs and some of the tensions that arose as we planned for first graders to recognize the power of their voice in connection to Martin Luther King Jr.'s legacy of peaceful protest. Second, the CRMT2 helped us to focus not just on the content of the lesson but also the importance of *how we enact* a lesson. Through our coaching partnership lens, we describe some of the instructional decisions we made for facilitation in reconstructing the district lesson plan.

The District-Created Lesson and the Mathematics

The district lesson helped us to better understand the mathematics that the kids would need to access a deep level of thinking around estimation and magnitude. Common Core Standards for Mathematics (National Governors Association Center for Best Practices & Council of Chief State School Officers, 2010) detail that Standard Mathematical Practice 5, "Use Appropriate Tools Strategically," includes students' abilities to "detect possible errors by strategically using estimation and other mathematical knowledge" (p. 7). Further, the standards state that "through activities that build number sense, they understand the order of the counting numbers and their relative magnitudes" (p. 13), which is one of the key areas of mathematics learning in Grade 1.

To explore the magnitude of numbers, the district lesson provided a mix of images that would support students in making both large and small estimations. It also used images of objects in various sizes (bicycles vs. small stars). This allowed the students to focus on value and not the space the images were taking up. The lesson mostly focused on cartoon images but did include a few real-world images such as part of a brick wall and a gumball machine, showing kids how estimation might be beneficial in a real-world experience. The lesson also helped us to focus on estimation as relational understanding, by asking questions such as *When does two feel big? When does two feel small?* These types of questions can help children to consider how magnitude might change based on the context, such as two years compared to two seconds. Such questions helped us set the stage for our understanding of estimation and magnitude and were the foundation to our planning with the CRMT2 Framework.

Transformation 1: A Focus on Power and Humanization

Two dimensions of the framework stood out to us the most as we began planning because they were newer to our thinking about planning mathematics lessons. First, *Analyzing and Taking Action*. As teachers, this dimension felt mostly absent from our mathematics instruction thus far in the year. We wondered, what does *Analyzing and Taking Action* look like for first-grade children? How is this connected to estimation and magnitude? The Social Justice Standards (Learning for Justice, 2022) helped us to consider some ways that we might change the content of the lesson to include social critique and action. The following two standards guided our planning:

1. Action 16: I care about those who are treated unfairly.
2. Action 19: I will speak up or do something if people are being unfair, even if my friends do not.

We wondered about a six-year-old's understanding of what it means to speak up and about their experiences of being treated unfairly or seeing others treated unfairly. This led us to think about protesting, which was also connected to our first-grade Social Studies standard of learning about Martin Luther King Jr. How do our students understand protest and Martin Luther King Jr.'s influence of peaceful protest? And importantly, how can we connect this Social Studies standard to the estimation and magnitude concepts in the mathematics standards? Olivia had the idea of using protest images of people, both historical and current, to apply the same mathematical thinking skills that would be needed in the district lesson. Figure 6.1 shows an example of how we transformed the type of image students used for exploring magnitude. Later in the chapter, you will see how students employed a variety of strategies to estimate the number of people at protests in various images with crowds of people attending in protest. These protests ranged from having crowds of smaller magnitude to those with much larger groups of people, allowing for students to engage in making sense of quantity and number in a variety of ways connected to real-world images of protests.

FIGURE 6.1 A Comparison of What the Original Problem Was Like (Paraphrased) and the New Problem From the Transformed Lesson

RE-CREATED DEPICTION OF IMAGE FROM DISTRICT LESSON

About how many frogs are there?

2, 20, or 200?

IMAGE IN TRANSFORMED LESSON

About how many people are there?

SOURCE: Frog icon by Gurrzzza/istock.com; Demonstration 12F in Venezuela 2014 by Wilfredor, commons wikimedia. org, CC0 1.0 Universal Public Domain Dedication, https://commons.wikimedia.org/wiki/File:Demostration_12F_in_ Venezuela_2014.jpg

With protesting as our context for the task, we wanted to help students understand several things. First, why do people protest? And, what does peaceful protest mean? Importantly, we wanted students to understand that they, even as children, can take action and the ways in which they can do so. We wanted to help them understand that Martin Luther King Jr.'s influence on peaceful protest still stands strong today as a way to stand up against injustice.

We quickly realized students needed more background knowledge before jumping into estimating within images of protests, if we wanted them to truly *Analyze and Take Action*. How can our students understand the history of protest without thinking deeply about the past of our country and the complexities of injustice? Specifically, how do you understand Martin Luther King Jr.'s protesting for civil rights without first understanding the systemic oppression and stronghold of slavery that still impacts society today? It was eye-opening in our planning to realize what it takes to help children see their place in current social issues tied to the evolution of protesting for civil rights over the course of history. We realized that this could not be done in one lesson. Thus, our lesson resulted in some important pre- and post-work surrounding the mathematics lesson on estimation.

Prior to the lesson, students learned about Martin Luther King Jr.'s legacy through reading stories and watching short videos, focusing on the meaning of protest and how he inspired others to take action against racism. We partnered this with several other activities to help children recognize unfairness around them. We asked them to reflect on and share to the following prompts:

1. *What is a rule that you think is unfair?*

2. *When have you or anyone you know been treated unfairly?*

3. *Do you think kids and adults can both change unfair rules or laws?*

These prompts elicited some important ideas about "fairness" through the eyes of a six-year-old. For example, students noted that school rules often applied to children but not to teachers, such as talking in the hallway. Young children recognized the power dynamic of kid versus adult in the school setting. They also recognized violence as unfairness, drawing on stories of peers or friends reacting through physical violence when feeling angry. We were able to connect their personal stories of unfairness to the injustices that Martin Luther King Jr. and other civil rights advocates continue to fight against. The final question about adults and children having the ability to change rules or laws elicited interesting conversations around students' understanding of power dynamics of adults and their place in society. Many students felt that presidents and principals can change laws and rules whereas other people do not have the ability to create these types of changes. The students also underestimated their own strengths as a community to make positive changes within their world and had a clear belief that kids did not have the ability to make any type of changes.

We wanted to bring light to how people, even children, can use their voice to make change in situations of injustice, tying back to how *Taking Action* is an important part of their civil rights. We read the book *Say Something*!

by Peter Reynolds (2019), which illustrates the many ways children can use their voice, and most of all, why their voice is needed.

We began the estimation and magnitude lesson by changing the numeracy routine offered by the district to focus on how protest may have evolved over time but is still a crucial part of society today. The district-created lesson's numeracy routine had students consider an estimation clipboard (Wyborney, 2023) image, estimating how many marbles were in a glass jar before "revealing" the actual number of marbles. While this routine is a powerful one for developing estimation sense, we wanted to elicit ideas of estimation and magnitude in a way that could also allow first graders the opportunity to critique the world around them. We changed this to a routine that recognized how two images were the same or different, showing Martin Luther King Jr.'s March on Washington for Jobs and Freedom protest compared to a recent Black Lives Matter protest in the same location, which you can see in Figure 6.2.

Let's take a closer look at how students engaged in the launch of the task with this purposeful adaptation.

A Closer Look: Task Launch

Children sit as a whole-class community on the carpet at the beginning of the mathematics lesson. Olivia, the teacher, projects the two images, one of the Washington, DC, Martin Luther King Jr.'s protest (1963) and another of a Black Lives Matter protest in the same spot right in front of the Washington Monument by the Lincoln Memorial Reflecting Pool (2020). The first graders peer curiously at the images, and Olivia asks, "What is the same about these images? What is different?" After a moment of thinking time, nearly every hand raises in anticipation of sharing their ideas.

While students began the same and different routine by noticing the "oldness" and "newness" of the pictures side-by-side, we noticed that their observations became more focused on the mathematics and larger social implications as they thought further.

Evelyn:	*I think there are more people in the black and white one.*
Olivia:	*Do you all think these are small protests or big protests?*
Tashawn:	*(Points to the Martin Luther King Jr. protest photo) I see people all the way back there! I feel bad for those people.*
Olivia:	*Why?*
Tashawn:	*Because they are so far away!*
Olivia:	*Remember in the book we read? What were people in the very back of the protest doing?*
Mathias:	*Sitting on each other's shoulders.*

FIGURE 6.2 Numeracy Routine

DISTRICT SUGGESTED NUMERACY ROUTINE

TRANSFORMED LESSON NUMERACY ROUTINE

How are these pictures the same? How are they different?

SOURCE: Image of glass beads in a bottle by vladj55/istock.com; Martin Luther King Jr. image by Getty Images; Protest image by LeoPatrizi/istock.com

NOTE: In the transformed numeracy routine, the image of Dr. Martin Luther King Jr. was in black and white; the image of the Black Lives Matter protest was in color.

Olivia:	*Yes, and climbing up trees. And, most importantly, they all were being so quiet so they could try and hear. It must have been really important.*
Kaiden:	*Both of them have a lot of people.*
Olivia:	*They both do have a lot of people. And I heard Evelyn say that one is smaller. Which one is smaller?*
Evelyn:	*The colorful one.*
Olivia:	*It might be smaller, but is it a small protest?*

Holly:	*I am wondering more about the noticing that one of the pictures is old and one of the protests is not. What does that tell us?*
Jada:	*That that one was a long, long, long time ago (points to the Martin Luther King Jr. photo) and that one is right now (points to Black Lives Matter photo).*
Holly:	*Yeah, are people still protesting today?*
Class:	*Yes.*
Holly:	*So, one is a long time ago, and one is today. Maybe not right now, today, but very recently.*
Yusif:	*Then why does the older one have more people and the one today has less people?*
Mathias:	*Because it's a smaller picture, and they aren't showing the other people.*
Olivia:	*Yes, so the black and white picture is taken from higher up, at a different angle, so that's why we can see the full Washington Monument. So, we can see all the people back there.*
Olivia:	*Do you think these protests are for similar things? Different things?*
RJ:	*They are for the same thing.*
Holly:	*Why? How do you know that?*
RJ:	*Because there were lots of different protests for Brown people.*
Holly:	*So, people are still protesting today for Black and Brown people and their rights? Is that what you are saying? That you have heard of those protests?*
RJ:	*(Nods yes.)*

One student goes on to notice that there is a Black Lives Matter flag and recognize it from the White House lawn from their recent visit to Washington, DC.

PAUSE & REFLECT

- What do you notice about the interactions between students?

- In what ways is the mathematics discussed?

- In what ways are the social issues discussed?

While our adaptation of the Same But Different routine (Looney Math Consulting, 2022) gave students the opportunities to have mathematical noticings and wonderings considering the magnitude and distribution of the people in the crowds, they also were able to have social noticings and wonderings. Chiefly, first-grade students noticed that people are still protesting in this location today for the same rights of Black and Brown people that Martin Luther King Jr. protested for years ago. Students also connected to the Black Lives Matter protest, noting that they had seen similar protests on the news or in local community media coverage (see Figure 6.3).

FIGURE 6.3 Transformed Lesson Numeracy Routine Poster

The poster captures the same and different conversation that we had as a class during the numeracy routine.

To bring the ideas about protesting back around after the lesson on estimation and magnitude, we also had students engage in a writing prompt reflecting on how they can use their voice when they see injustice (see Figure 6.4). Students drew on their new knowledge of protest and how it can impact change, writing about how they can participate in speaking up against injustice.

Another student, as seen in Figure 6.5, suggested protesting for a specific cause, the right for kids to vote.

Viewing this lesson through the dimension of *Analyze and Take Action* first pushed us to really consider the ways in which first-grade students might make sense of the world around them and also helped us to make important interdisciplinary connections between mathematics and social studies. By recognizing that this dimension is the one that is most easily lost in our mathematics lesson planning, we were able to move it from the margins and center it in our planning while adapting this lesson.

Transformation 2: A Focus on Who We Are

The second dimension of the CRMT that we focused on for the content of our lesson was *Cultural and Community Funds of Knowledge*. We wanted to help the students who have familial connections outside of the United States make those personal connections to their lives

FIGURE 6.4 Two Student Work Samples

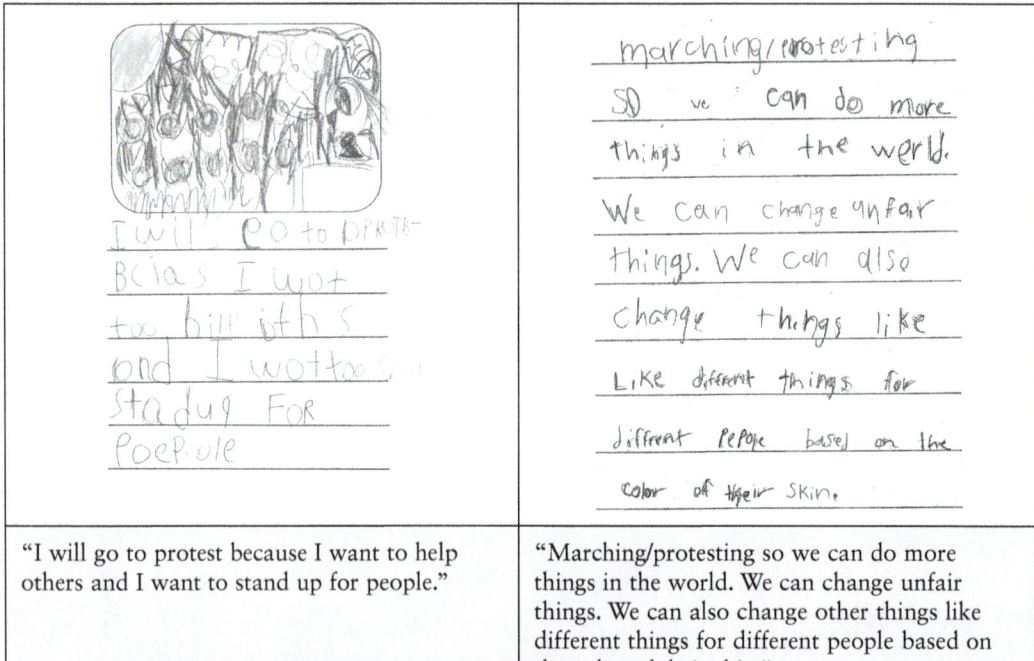

| "I will go to protest because I want to help others and I want to stand up for people." | "Marching/protesting so we can do more things in the world. We can change unfair things. We can also change other things like different things for different people based on the color of their skin." |

These student work sample include a drawing of people attending a protest and sentences another student wrote about protesting.

FIGURE 6.5 Another Student's Work Sample

| "We can protest. Kids can vote." |

This student work sample includes their picture and sentences they wrote about protesting.

throughout the lesson, while still remaining authentic to the focus issue of people protesting for rights. Further, we felt this dimension helped us to remember that mathematics can be a window for students who may not have those connections outside of the United States to see, learn, and recognize that people all across the world peacefully protest when they see an injustice. To expand their experiences beyond what they know and see in America, we were intentional in providing both global and local examples. We found protest images from Sri Lanka and Ethiopia, countries that were representative of several of our students' families. As the students were given the various images, one student immediately recognized the flag from Sri Lanka (see Figure 6.6) and excitedly shared it with her table. Additionally, with our multilingual students speaking mostly Spanish at home, we were intentional in including

FIGURE 6.6 Two Images of People Protesting

Two images of people protesting that we chose to use, and which are representative of children in this class in different ways.

SOURCE: Anti-government protest in Sri Lanka by AntanO, commons wikimedia.org, CC BY-SA 4.0, https://creativecommons.org/licenses/by-sa/4.0/deed.en; Protest image by PeopleImages/istock.com

protests with signs written in Spanish. Lastly, we included a protest that was local to our school district from a recent high school walkout. We chose to pick more current protest images that which students may have seen in local news or participated in themselves. Figure 6.6 shows some of the images we used purposefully to connect to our students and their lived experiences.

Transformation 3: Rethinking Lesson Structures

While the focus on power and students' funds of knowledge had us considering what content we needed to adapt in the lesson, other dimensions of the CRMT2 supported us in grappling with instructional decisions for the enactment of the lesson. As it was written, the district lesson said very little about the teacher moves and specifics for how to enact the lesson, besides specifying a "whole group" section and a "math workshop" section that included station rotations with different activities for several groups, including a teacher center. This given structure led us to consider the dimension **Disrupting Status and Power**. Ultimately, the CRMT2 helped us to understand that *how we enact a mathematics lesson matters just as much as the content that is in our mathematics lesson*. We know that math classes in schools all around the country are still tracking kids based on assumed ability. In our lesson, we wanted to make sure not to separate the students by perceived understanding and skill, as is sometimes done with the math workshop model. Rather, we co-planned a lesson that was task-based and opted for a structure better suited to allowing all students equitable access to the mathematics, where they could showcase their different strengths through multiple entry points. This was our way of using the CRMT2 for a structure that allowed children (and us) more agency. As we considered which classroom structure would best **Disrupt Status and**

Power, we recognized how intricately connected the CRMT2 strands are to one another and how our planning decisions within one dimension greatly impacted another.

The focus on **Disrupting Status and Power** helped us consider the Cognitive Demand of the lesson and how we would *Scaffold Up*. Figure 6.7 shows the original suggestion and the task we created. The cognitive demand of the task remained rigorous, as we challenged the students to work in small groups to show and explain a strategy for choosing their estimate. Student groups were created by focusing on diverse strengths to provide opportunities to hear different perspectives and learn from one another. Each group had at least one picture of a small protest, one picture of a midsized protest, and one picture of a large protest. One way that we provided access and scaffolded up was through the offering of manipulatives. We printed the pictures of the protests, put them in pocket sleeves, and offered dry-erase markers, unifix cubes, and hundreds charts for students to choose from if they wanted a tool to support their thinking. Each protest picture had choice cards with reasonable estimates of magnitude

FIGURE 6.7 Comparison of the District's Suggestion for the Lesson and the Replacement Task

DISTRICT SUGGESTIONS FOR STATIONS	NEW TASK
 Station & Small Group Ideas Choose from the following: • Choose a game from an earlier unit. • Choose 1–2 activities for differentiation from an earlier unit. • Use an online learning program. • Choose a lesson from this unit to use with a small group. • Choose a number-sense game.	We can make a reasonable guess about how many there are without counting. This is called an estimation.  What does *reasonable* mean? What is a number of people in this picture that would be too low? Too high? How do you know? Today we are going to use tools to think about *how many* people we think are in the pictures from protests around the would. You should be able to explain to others *why* your estimate makes sense. Tools: • Picture • Expo markers • Hundreds chart • Unifix cubes  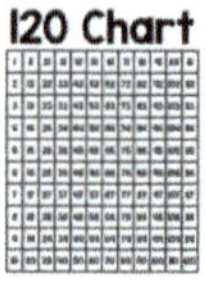

SOURCE: Martin Luther King Jr. Image by GettyImages; Markers image by Marco di Bello/istock.com involving three quantities as described in the state standards: a one-digit numeral, a two-digit numeral, and a three-digit numeral (e.g., 1, 10, 100) to choose from once they were ready to make a final estimate.

In Figure 6.8, we highlight how a partnership used two different student ideas to explore separate photos of protest. The first student work sample reveals how a smaller-sized protest acted as an access point into the task, allowing the student to be elevated in their thinking as they counted by 1s. The partnership then chose a medium-sized protest that lent itself to the strategy of making groups of estimated tens, which extended from their thinking of using groups of one in with the first image. If students had been broken up into perceived ability groups, many students wouldn't have seen all the various sizes of the protests and would not have had the opportunity to be challenged or learn alongside their peers. It was essential for us to **Disrupt Status and Power** through this lesson, not only for children and their right to learn and love mathematics deeply, but also for ourselves to confront our own bias in everyday math lessons by not grouping students on what we deem they can or cannot do.

FIGURE 6.8 Two Student Work Samples Exploring Ideas of Protest

SOURCE: Partner 1 image from Indiana University (2023). Partner 2 image from Ashford (2022).

To **Affirm Multilingualism** of our students, we provided the space for students to work with a language buddy, communicating in their home language, and we uplifted Spanish-speaking culture through the pictures we chose. Through all of this and the task-based structure, we aimed to **(Re)Humanize Mathematics** for children, broadening notions of what counts as mathematical knowledge. The goal was not just to simply choose the best estimate but instead to reason with the tools and images in creative ways for determining magnitude, as seen here through the richness of **Building on Student Thinking and Ideas**.

Let's take a look into a snapshot of how the task unfolded with this adapted structure.

A Closer Look: Task Enactment

Students excitedly get to work on looking at their images. One group grabs the hundreds chart, others begin chatting about groups of people they are seeing in the photo. Several children begin circling and writing using the dry erase markers but then erase as they discuss and refine their thinking. Olivia and Holly walk from group to group, just watching to start. Holly listens in on one partnership, curious to see how they are making sense of the varying sized circles on their paper:

Isabelle:	*So, this has to be like 60. (circles group on the image that already has varying sized groups circled).*
Holly:	*How are you all using these groups to make an estimate of how many people you think are at the protest?*
Evelyn:	*It will be easier to count if we make groups of about ten.*
Isabelle:	*We could erase these smaller groups and make bigger groups.*
Evelyn:	*Good idea. One of the choices is 900, so maybe we can find groups of 100.*

SOURCE: Ashford (2022).

To **Honor Student Thinking and Ideas** and **Distribute Intellectual Authority**, we made intentional pedagogical decisions to act as facilitators of connecting big mathematical ideas and supporting communication. The strategies, representations, and solutions rested in the hands of our students, *each and every* student. Importantly, we preplanned this facilitation using specific teacher moves. Prior to the lesson, we committed to using three prompts adapted from the CRMT teacher moves:

1. *How does your group's model help us to understand the magnitude of the people shown from the protest?*
2. *How does _____'s idea help us to understand estimation?*
3. *Tell me more about your group's mathematical thinking.*

The groups continue to collaborate, excitement still at high level for making sense of magnitude in pictures of the protest. One group exclaims, "Hey! I saw this on the news!" from across the room. Another partnership mentions, "This one reminds me a lot of the Martin Luther King protest we saw in the book." We continue to be amazed at the dynamic ways students explore quantity and magnitude. Holly heads over to try out one of the prompts as a partnership discusses using a 120 chart.

SOURCE: Ashford (2022).

| Holly: | Tell me more about your group's mathematical thinking. |
| Justin: | So, there are about 5 people over in this area (circles a group on the picture of the protest). So, there are 40 groups, so let's see. Let's circle 40 (on the 120 chart). |

So, we already have 40. And so, we are pretty sure ... wait hold on let me think. I don't think it's going to be 40 (erases the circled 40 on the 120 chart). How many were there?

Tashawn: Counts the groups they circled.

Justin: Oh, there's five groups so far (counts by 5s on hundreds chart 5 times). Oh, so it's 25. So, we figured this one out, we think it's about 25 people. Let's see if 25 is an answer (pulls out choice cards).

Holly: So, you estimated and said that there are about 25 people. Which one of the choices would make the most sense?

Justin: (Pulls out choices of 2, 20, and 200). 25 is close to 20. So maybe it's about 20 people.

PAUSE AND REFLECT

- There are several vignettes highlighted throughout this chapter. In what ways were students positioned as knowers and doers of mathematics?

- How did the structures make space for multiple forms of knowledge ideas?

A TOOL FOR LESSON TRANSFORMATION

This chapter highlights how Holly and Olivia grew in their coaching partnership to strengthen a single mathematics lesson, transforming it into a more culturally responsive and inclusive experience for children. Along the way, they recognized the dual power in both the content of the lesson and the enactment of the lesson. Their main content changes included those that led to first graders *Analyzing and Taking Action* through mathematics. On the other hand, many of the CRMT2 dimensions illuminated the need for altering the pedagogical moves and classroom structures embedded within the lesson. In Figure 6.9, Holly and Olivia summarize their reflection on the district lesson through the dimensions of the CRMT2, as well as their planning notes for changes they made when prioritizing their children as knowers and doers of mathematics.

FIGURE 6.9. Holly and Olivia's Reflections on Using the CRMT2

KNOWLEDGES & IDENTITIES DIMENSION	DISTRICT LESSON NOTES	ADAPTED PLANNING NOTES
Centering Cultural and Community Funds of Knowledge *How does my lesson help students connect mathematics with relevant/authentic issues or situations in their lives?*	Images for magnitude are cartoons, such as stars or hearts. Two of the images are of real-world pictures (gumballs and a brick wall).	• Images for magnitude will include global protests: ○ Include protests from the countries of our students (i.e., Sri Lanka) ○ Include local protests (district high school walkout) ○ Include Spanish-speaking protesters
(Re)Humanizing *How does my lesson support creativity, broaden what counts as mathematical knowledge, and affirm positive math identities for all students?*	The whole group lesson is interpreted as students sharing out one at a time. Small group stations are suggested where students work on different games, a computer program, and a small group lesson.	• Use a whole group task structure. • Students work in small groups and are offered tools to support a variety of creative strategies for estimating. • Tools will include images in sleeves with expo markers, hundreds charts, and unifix cubes • Teacher role will be to ask advancing and assessing questions, remain neutral, and honor the thinking of the groups.
Honoring Student Thinking and Ideas *How does my lesson create opportunities to elicit, express, and build on student mathematical thinking in multiple ways? (e.g., through gesture, pictures, words, symbols)*	This is unclear in the district lesson. Possibly, students could share with the whole class some strategies and the teacher could annotate on the projected images.	• Using the tools and small group thinking, they can use multiple entry points to show their mathematical models and explain their thinking.
RIGOR & SUPPORT DIMENSION	DISTRICT LESSON NOTES	ADAPTED PLANNING NOTES
Sustaining Cognitive Demand *How does my lesson enable <u>all</u> my students to closely explore and analyze math concepts(s), procedure(s), and problem-solving/ reasoning strategies?*	Images are available to visualize size, but it is unclear how all students would closely analyze math concepts. In the small group, would they be ability grouped and so some would have higher cognitive demand than others?	• Task structured with heterogenous grouping. • All students are given access to the same tools. • Extension planned and available for all students: give an image of a larger protest (in the thousands) and have them create an estimate.
Scaffolding Up *How does my lesson maintain high rigor with high support for all students?*	This is unclear, and we wonder about the purpose of the small group rotations and whether that would maintain high rigor and high support for all students.	• Tools and visuals for access • Peer-to-peer discourse • Teacher facilitators with preplanned advancing and assessing questions

(Continued)

FIGURE 6.9. *(Continued)*

RIGOR & SUPPORT DIMENSION	DISTRICT LESSON NOTES	ADAPTED PLANNING NOTES
Affirming Multilingualism *How does my lesson make space for multilingual learners (MLL) to be central participants in mathematics activities?*	This is not specified in the lesson.	• Students are always welcome to speak their home language with a language buddy. • Include protests with signs in other languages. • Highlight the expertise and contributions of our MLL for the class.

POWER & PARTICIPATION DIMENSION	DISTRICT LESSON NOTES	ADAPTED PLANNING NOTES
Distributing Intellectual Authority *How does my lesson distribute mathematics authority and make space for multiple forms of knowledge and communication?*	This is unclear in the district lesson. Possibly, students could share with the whole class some strategies and the teacher could annotate on the projected images. Depending on the structure, students could turn and talk before sharing.	• Using the tools and small group thinking, they can use multiple entry points to show their mathematical models and explain their thinking. • Students will focus on the skill "communicator" with explicit attention to how we listen to our peers and their ideas.
Disrupting Status and Power *How does my lesson disrupt status differences, entrenched stereotypes, and inequitable power relationships present in all mathematics classrooms?*	This is not specified in the lesson. The small group rotations make us wonder if it would be perpetuating entrenched stereotypes with homogeneous grouping of students.	• Through the task structure, all students will be given the same opportunities to think deeply about the mathematics, regardless of known stereotypes of who "can" and "can't" do mathematics. • Heterogenous grouping with discourse structures will honor all students as knowers and thinkers of mathematics.
Analyzing and Taking Action *How does my lesson support student use of mathematics to analyze, critique, and address power relationships and injustice in their lives?*	Not present in the lesson	• Images of protests for estimating and magnitude • Interdisciplinary connection to Social Studies curriculum and Martin Luther King Jr. (peaceful protest) • Prework around how students can use their voice in protest, what protest means, why people protest • Prework with students brainstorming spaces where they recognize unfair rules or people being treated unfairly • Postwork where students consider what they have learned about peaceful protest and write about how they might use their voice to stand up for something that is unjust.

DISCUSSION QUESTIONS

- Consider the makeup of your classroom. How might you adapt your next lesson to uplift students' cultures and funds of knowledge?

- If you were implementing this lesson, how else might you extend student learning to promote *Analyzing and Taking Action*?

- How might you use this tool as a team or in a critical learning partnership to adapt a textbook or district-created lesson?

CHAPTER 7

USING CRMT TO TRANSFORM A FOURTH-GRADE MATHEMATICS CLASSROOM OVER TIME

By Holly Tate and Kaitlin Kaplewicz

This chapter highlights the learning that unfolds over several years between mathematics coach Holly Tate (the same coach from Chapter 6) and fourth-grade teacher Kaitlin Kaplewicz. Together the partnership explores a series of goals to become more culturally responsive through cycles of planning, implementing, and reflecting. The Culturally Responsive Mathematics Teaching Tool (CRMT2) plays a critical role in this continual process, as Holly and Kaitlin use the tool to make decisions and reflect retroactively on their journey. Kaitlin and Holly are white mathematics educators who work in urban classrooms, and this chapter includes their experiences within two different schools. Let's learn from the chronological progression of their learning in the pages that follow as we hear their story.

BECOMING CULTURALLY RESPONSIVE MATHEMATICS TEACHERS

Enter a school that has been deemed as "failing." Standardized test scores have led to a status that does not meet accreditation standards of the state. Within the community, many turn up their noses at the idea of sending their kids to "that school." Those who can afford it opt for private school. District and state-level administrators are often in the building, looking for the next "fix" to the problem of test scores. "What will it take? What can we do? What is wrong with how things are being done?" are questions often asked amongst the groups of people leading the charge for change. Yet, the same things are done. Drill students, test students, sort students, drill students … the cycle repeats.

We begin together as a coach and teacher partnership, quickly falling into step with how things "have always been done" in the building. We both quickly recognize that it's not working. Students are not excited for mathematics. They often say things like "I hate math" or "I'm not good at math" We know something needs to change, but what? How can we make this a space where students love mathematics and understand their mathematical power and identity?

When we began our coaching partnership, neither one of us could imagine how learning together about Culturally Responsive Mathematics Teaching (CRMT) would forever change our minds and hearts. In this chapter, we describe how we used the CRMT2 in our learning partnership over a time period of five years, focusing on different dimensions throughout our learning. We highlight several cases and examples that stand out to us as monumental moments of our learning. The CRMT2 helped us to take new risks and led to tensions and reflections that transformed us as mathematics educators and transformed the classroom experience of learning mathematics for our students.

YEARS 1 THROUGH 3: A FOCUS ON DISRUPTING STATUS AND POWER

A demographics snapshot into the school where our story begins reveals a self-identified racial make-up of 50 percent Black, 23 percent white, and 20 percent Hispanic. Around 85 percent of students are impacted by poverty, 20 percent of students are multilingual learners, and 20 percent are receiving special education services. The stereotypes of students who *could* and students who *could not* succeed were pervasive in our school and intricately embedded into each of the systems at play within the walls of our classrooms. During the first year of our coaching partnership, we spent a lot of time thinking about ways to disrupt inequities. This disruption moved beyond looking solely at a lesson, instead focusing on the daily structures, the role of the teacher, and our positioning of students as we engaged in mathematics.

We thought about the dimension of the CRMT2, *Disrupting Status and Power.* But we wondered, How *do* we disrupt the status and power relationships that our school, communities, families, and students are so accustomed to?

We recognized a powerful relationship among three key elements of the CRMT2 lesson analysis tool that helped us focus on our work to *Disrupt Status and Power: Honoring Student Thinking and Ideas, Sustaining High Cognitive Demand,* and *Scaffolding Up* (Figure 7.1).

FIGURE 7.1 Interwoven Strands, Focus 1

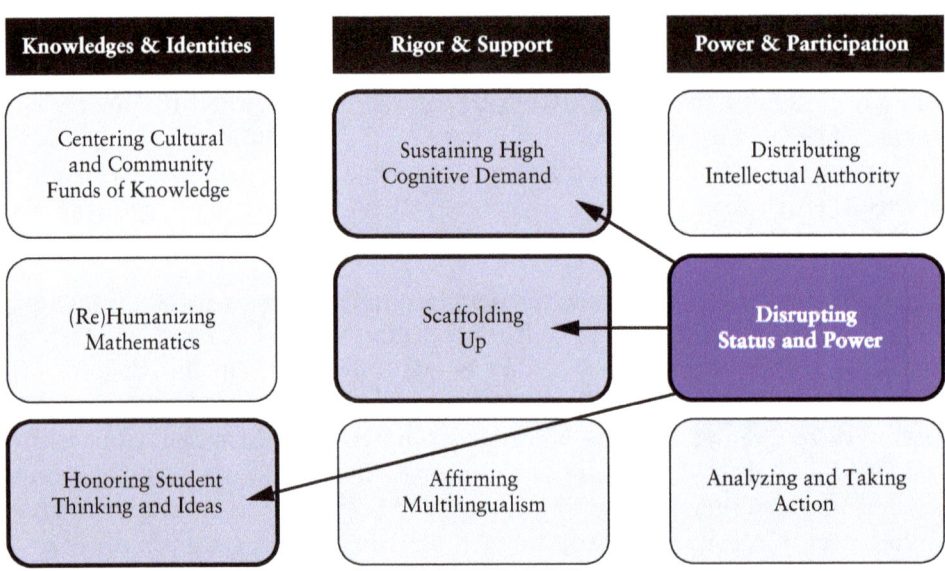

We knew our very first goal for learning as teachers was to transform our classroom structure for mathematics. To start the school year, all students had been put into fixed-ability groups based on a prescreening, standardized assessment. Our "high" group received rigorous task-based instruction, while our "low" group received "I do, we do, you do" instruction. Further, those groups deemed "low" were mostly made up of neurodiverse and multilingual students, and most of the "high" groups were made up of affluent, white students. In order to disrupt the stereotypes that were deeply entrenched in the ability-grouped structure, we needed to drastically change our classroom structure for rigorous mathematics that honored unique student thinking and ideas, while also questioning the systems at play at the larger school level: How were educator and leadership bias present in these groups? How were our systems and classroom structures perpetuating inequities based on stereotypes of who is and is not worthy and capable of rigorous mathematical experiences? We decided to try our hand at some task-based structures with diverse-strength groups, in which students would collaborate together. Our collaborative groups were formed based on student mathematical strategies, language partners, or randomly, instead of by perceived ability.

We sit at a horseshoe table, a preplanning coaching meeting. Holly notes, "This is going to be a big change. Do you think you would be open to trying it? It might be messy. It might be hard. It might not work." Kaitlin agrees and is excited for the change. *"What we are doing now is not working; it is time to be better for students."* We dive into a unit plan for multiplication and division, wondering, *"What will this look like over time?"* and *"How can we honor student thinking and ideas?"*

Though we felt some apprehension in attempting something new, we recognized the powerful implications of what this could mean for student identity. As a coaching partnership, we agreed to try a task structure, based on the Five Practices for Orchestrating Productive Mathematics Discussions (Smith & Stein, 2018), with heterogeneous groups every day for an entire unit. With this task structure, we launched a task for the entire group, then provided students the opportunity to productively struggle with the mathematics in their collaborative groups. As students were working, we monitored their thinking and ideas, looking for strategies and big mathematical concepts while asking purposeful questions to assess student understanding or extend their thinking. At the end of the task, we went back to the carpet as a whole class and made connections over several purposefully selected group strategies. Over the course of a few weeks, we consciously positioned students to make meaning of larger equal group problems. The following vignettes highlight some of the pivotal student conversations that helped us to learn as we reflected on becoming a Culturally Responsive Mathematics classroom. Figure 7.2 summarizes how our initial changes connect to the CRMT2.

FIGURE 7.2 Reflection on Learning Within the CRMT2, Focus 1

KNOWLEDGES & IDENTITIES DIMENSION	ESSENTIAL QUESTION	PLANNING NOTES
Honoring Student Thinking & Ideas	*How do we create opportunities to elicit, express, and build on student mathematical thinking in multiple ways? (e.g., through gesture, pictures, words, symbols)*	• Use open-ended multiplication/division problems that are connected to area and equal groups. • What types of problems elicit area models without us telling students to create area models? • What are all of the ways students might solve these problems? • How do our anticipated problems connect to learning trajectories? • We can honor student thinking and ideas without inserting our own strategies as teachers. Students can learn from one another. • What ideas will we highlight over time? How do we create the space for all students' ideas to be honored?

(Continued)

FIGURE 7.2 *(Continued)*

RIGOR & SUPPORT DIMENSION	ESSENTIAL QUESTION	PLANNING NOTES
Sustaining High Cognitive Demand	*How do we enable <u>all</u> our students to closely explore and analyze math concepts(s), procedure(s), and problem-solving/ reasoning strategies?*	• The same task problem will be launched with all students. • Students will have a few minutes of independent think-time, then we will purposefully partner them with someone else based on their strategy or thinking. • What types of questions can we ask students so as to not take away the cognitive demand? • How do we remain open so as to not project our own ideas and instead elevate the ideas of the students?

Scaffolding Up	*How does my lesson maintain high rigor with high support for all students?*	• Tools that we will offer for access to students:
		  ○ Launching the task in a way that helps students to activate prior knowledge and understand the task
		  ○ Sentence frames to support student-to-student discourse
		• What questions can we ask to help students get started that do not take away their opportunity to think deeply about mathematics?

Zoom In: Cookies at a Bake Sale

COOKIES AT A BAKE SALE TASK

Marcus was baking cookies for our bake sale. He is putting 74 cookies each into 8 plastic containers. How many cookies will he need to bake?

With some experience in this new task structure (Smith & Stein, 2018) for learning mathematics, students were eager to get started working alongside one another. We notice the fourth graders grabbing the tools they want to show their thinking, such as base-ten blocks and unifix cubes, or using a model or a drawing to show their understanding of the problem. We walk around to the students, taking note of what they are doing. We often confer with one another to make sense of what the children are doing in the moment and to help each other make instructional decisions. We visited Kidra, who was finishing drawing her model of base-ten blocks with her partner:

    Kaitlin: *Kidra, can you explain how you thought about this problem?*

    Kidra: *I knew that there were 8, and I could put 74 in each to figure out how many cookies. I used these lines (points to paper) to make tens and the dots are ones. I counted all of the tens together and got 560. Then I counted all of the ones together and got 32. Then I put those two numbers together to see how many cookies.*

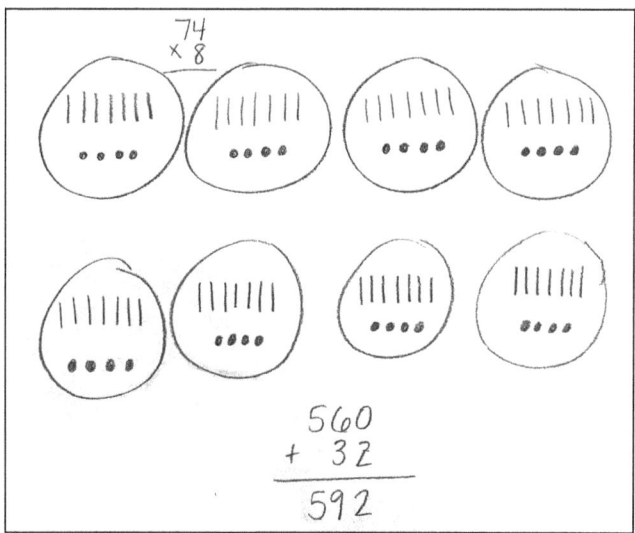

As a co-teaching partnership, we noticed that there was an important idea in how Kidra made sense of her total with her drawn base-ten blocks. First, she counted the tens and found that total, and then she counted the ones. This, we recognized, was early understanding of the distributive property. We knew Kidra's idea would be a brilliant connection for the class to consider.

We then went to listen to a partnership who was discussing two different solutions to the problem. One student, Amaya, had started with 74 and had divided them into groups. The other child, Marjorie, had created a model similar to Kidra's:

Marjorie:	*It doesn't say "equally." Yesterday when we did division, it said "equally."* (Refers to her strategy in her math notebook from the day before and shows her partner, then re-reads the problem). *"If 96 cookies are shared equally between the friends, how many will each friend get?"*
Amaya:	*Okay, I know what you are trying to say but it says if 74 cookies are in EACH—that was my key word.*
Marjorie:	*(Nods, thinking for a minute).*
Amaya:	*This is a hard choice.*
Marjorie:	*It is a hard choice.*
Holly:	*Could you all use your pictures? Could you talk about the pictures you drew and which might make the most sense with the story? What is happening in your picture with cookies and containers, Amaya?*

Amaya:	I used tallies and 8 circles.
Holly:	And what do those 8 circles represent?
Amaya:	The 8 cookies. There are 8 cookies.
Holly:	So, each circle is a cookie? Those are the 8 cookies?
Amaya:	Yeah, and I put … wait. (Rereads the problem). Oh, at first I thought there were 74 cookies and 8 cookies. But I see that they will be in 8 containers.
Marjorie:	Is that where you messed up? It's okay, we all make mistakes. Look at how I put the cookies into the bins.

PAUSE AND REFLECT

As educators, we continue to wonder about the best ways to **scaffold up**. In this moment, it felt like asking for the students to consider their models was a way for them to work through their understanding of the context and how they got different solutions.

- What questions can we ask to support students in their mathematical argumentation?

- What is our role in positioning students as the ones in power of their mathematical thinking and ideas?

After students shared in partnerships, we asked them to join as a classroom community on the carpet to think more deeply about some of their peers' ideas. We thought it would be a pivotal moment for students to make connections between the different ways students were using the distributive property. Students looked at Kidra's model from her partnership, where she first counted the tens and then counted the ones, and compared this to a student who used multiplicative thinking through partial products. The class contemplated the 74 cookies and how they were broken up into 70 + 4 across the models. We asked students to consider how these strategies are the same. How are they different? How does Kidra's counting (shown on the left) help us to see the 70 × 8 = 560 that Khan and Josue represented in their area model (shown on the right)?

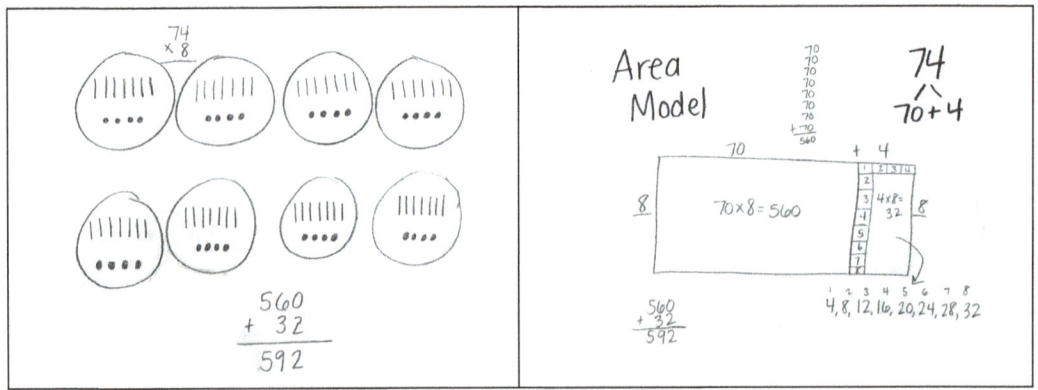

We often wondered as we were learning to *Disrupt Status and Power* how we would elevate the students who had previously described themselves as being "bad at math."

- How do we help students in this position to see their mathematical power?
- How do we provide the space for *all* students to be leaders and learners?

Using student work in a variety of ways as a whole class helped center all students in their brilliance. Over time, we developed an environment of confident math learners as they became more comfortable in asking questions, comparing ideas, and valuing all kinds of thinking. As we become more culturally responsive in *Disrupting Status and Power*, we begin to recognize the vast array of student thinking and how students are inventing strategies that allow their peers to understand the complexities of multiplication. This was very different from how we had previously taught equal groups problems, where we had explicitly taught each of the strategies that were deemed the "right" strategies. Instead, our students were the teachers. We began to understand that our role as teachers was to facilitate this open space where they could learn and grow from each other, a radical change that honored each student as a knower and doer of mathematics. As children engaged in various forms of models of multiplication, such as the area model mentioned in the vignette, our role as the teacher shifted to create the space for students to meaningfully use strategies that help make sense of the problem. We acted as facilitators in helping students to understand why these models made sense, how we could choose models based on the type of problem we are solving, and how the models are all connected. In Figure 7.3, we highlight some of the unique and brilliant contributions of our students and how our thinking was constantly pushed as we learned through the CRMT2.

Now, let's look at another example that showcases our initial work of *Disrupting Status and Power*, with particular focus on how systemic biases and labels are challenged.

Zoom in: Dog Man Books

DOG MAN BOOKS TASK

Our school just got a new shipment of Dog Man books. Each classroom will get an equal amount of books. If there were 84 books delivered to the school and there are 6 classrooms, how many new books will each classroom get?

We sit down to a coaching conversation after a lesson, in which we used the Dog Man Books task *to understand how students were making sense*

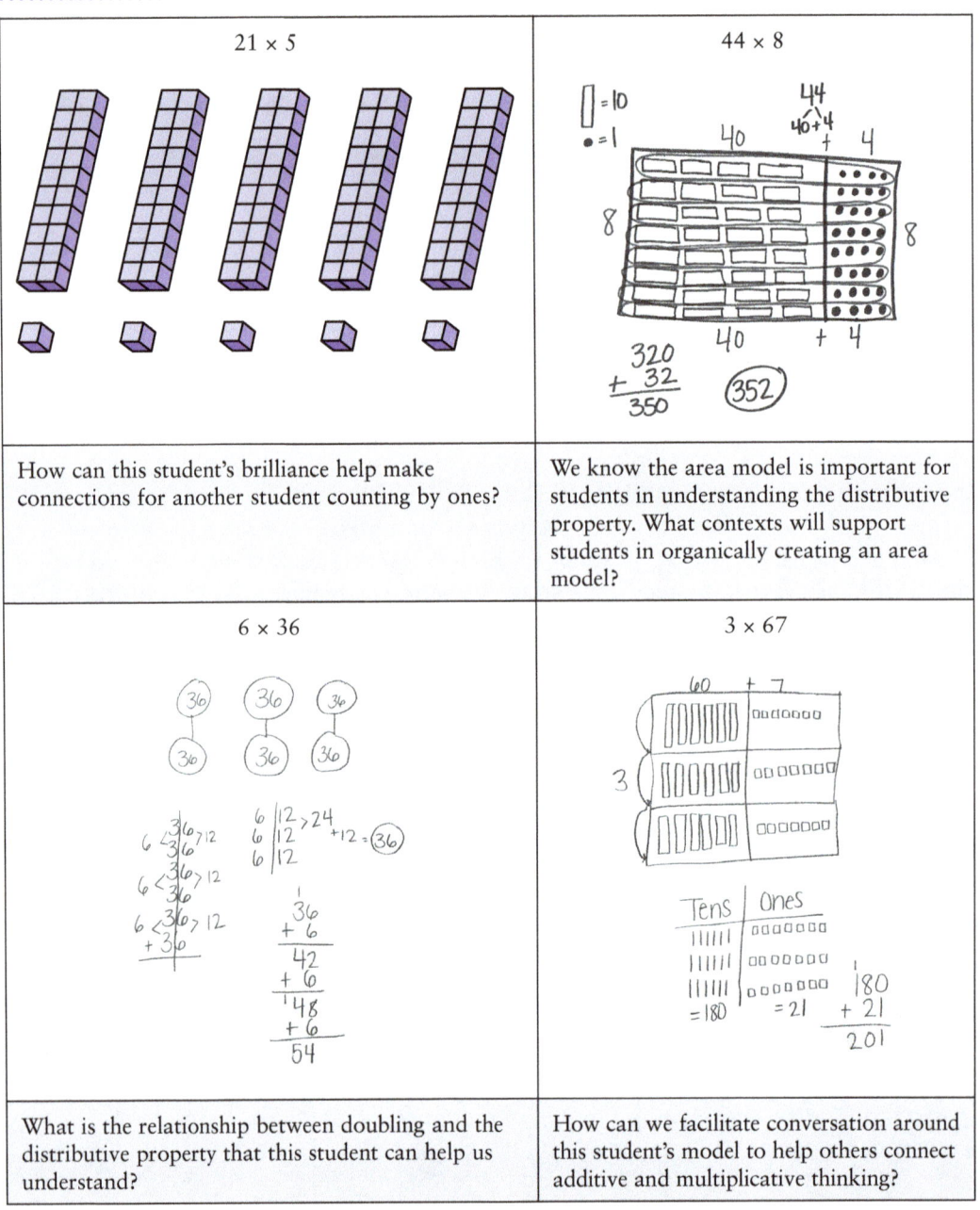

21 × 5	44 × 8
How can this student's brilliance help make connections for another student counting by ones?	We know the area model is important for students in understanding the distributive property. What contexts will support students in organically creating an area model?

6 × 36	3 × 67
What is the relationship between doubling and the distributive property that this student can help us understand?	How can we facilitate conversation around this student's model to help others connect additive and multiplicative thinking?

of a division context. Did they use equal grouping strategies, dealing out in groups? Did they recognize the relationship between multiplication and division? In what ways were students inventing procedures grounded in conceptual understanding? Through the task structure, we hoped to elicit student-invented strategies for division through representations and use of manipulatives. We had agreed at a previous session to capture some video of student groups working through the problem, so we could think through ways to support student-to-student discourse. We pulled up a video that includes a group consisting of one student dually identified as being neurodiverse and multilingual (Julian), a student being screened for Gifted education (Rayan), and a third multilingual student (Euan). We watch intently with excited anticipation as the kids explain their ideas to one another. About halfway through the group sharing, this interaction happens:

Julian:	*So what I did was . . . 42 and 42 is 84. So I broke it up. And 42 divided by 6 is 7. And I did it again, 42 divided by 6 is 7. And 7 + 7 is 14. So then what I did was, I got the 14 and multiplied it by 6 and it was 84. So, 84 divided by 6 is 14.*
Rayan:	*So, this is what you did. You broke up the 84 into 42 and 42. Then you knew 42 divided by 6 equals 7, 2 times. And 7 + 7 = 14. And then you did 14 × 6 = 84 to check it?*
Julian:	*Nods, yes.*
Rayan:	*Oh wow, that's good.*

We pause the video. "Wow," Holly comments. "Wow is right," Kaitlin adds. In this open space, Julian was in the position of mathematical competence as he shared his creative idea for dividing, while Rayan learned a new idea from him. The labels of "neurodiverse" and "gifted," which often lead to sorting of students and stratification of who "can" and "can't," were disrupted.

PAUSE AND REFLECT

- What surprises you or stands out to you as you read this interaction?

- In what ways might examples like this defy assumptions of who "can" and "cannot" do mathematics?

- How might the children in this vignette feel by having the opportunity to be in this space?

In reflecting on the video through the lens of the CRMT2 and considering what it meant to **Disrupt Power and Status,** this moment was a very proud one for us. In the space of this mathematics lesson, "typical" roles were reversed. We recognized the power in witnessing firsthand a student, who according to the systems that were in place previously would have been in a "low group," was positioned as a knower and doer of mathematics. His peer, who would have been in the "high group," reflected on Julian's idea and appreciated the brilliance behind it. We often wonder, what would we have missed out on if we hadn't changed the structure? Whose voices would we have heard, and whose voices would have been left behind? Who would have seen themselves as mathematicians, and who would we have failed because they might not have seen themselves this way?

YEAR 4: A FOCUS ON (RE)HUMANIZING

As time progressed, we realized we were ready to learn more deeply about becoming culturally responsive. We decided that, while we had a structure that we felt was *Disrupting Status and Power,* we still needed to work on *(Re)Humanizing Mathematics* in ways that were more connected to our students. This dimension stood out as a need more than ever to us as we continued our partnership in a new, incredibly diverse school. This school included a self-identified racial makeup of approximately 40 percent Hispanic, 33 percent Black, 16 percent white, and 8 percent Asian. About 83 percent of students were impacted by poverty, and 73 percent of students were multilingual learners. Languages of our students were an integral part of the teaching and learning in our building, in a district boast over 120 home languages, with a majority speaking Spanish, Amharic, and Arabic.

We continued to connect several different dimensions of CRMT and saw the impact that these dimensions had on students' mathematical identities.

In *(Re)Humanizing*, we spent a lot of time focusing on who our kids are, both inside and outside of the classroom. We recognize that we engaged in *Centering Cultural and Community Funds of Knowledge* and *Affirming Multilingualism* in our quest to *(Re)Humanize* the classroom, again using the intricately connected dimensions of the CRMT2, shown in Figure 7.4.

Within our new setting, it was clear there were stereotypes present about what students *could* and what students *could not* do; but this time, specifically, stereotypes were evident around our multilingual learners. Through

FIGURE 7.4 Interwoven Strands, Focus 2

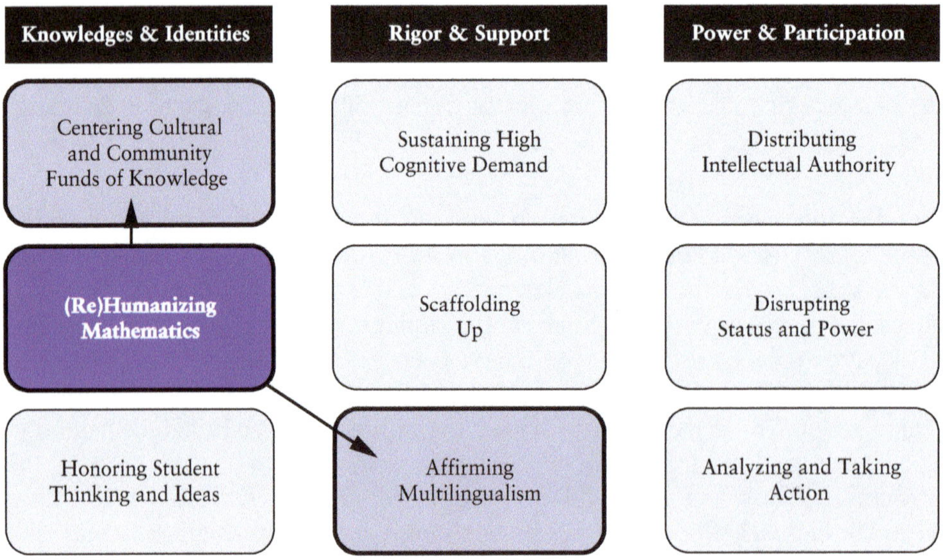

Knowledges & Identities	Rigor & Support	Power & Participation
Centering Cultural and Community Funds of Knowledge	Sustaining High Cognitive Demand	Distributing Intellectual Authority
(Re)Humanizing Mathematics	Scaffolding Up	Disrupting Status and Power
Honoring Student Thinking and Ideas	Affirming Multilingualism	Analyzing and Taking Action

our work with the components of the CRMT2 framework, we learned to broaden what counts as mathematical knowledge and ensure that as teachers, we were *Centering Funds of Knowledge* that each individual learner brought to our fourth-grade classroom, including positioning our multilingual learners as central participants in our mathematics community (*Affirming Multilingualism*). The main takeaways from this phase of our professional learning are highlighted in Figure 7.5.

FIGURE 7.5 Reflection on Learning Within the CRMT2, Focus 2

KNOWLEDGES & IDENTITIES DIMENSION	ESSENTIAL QUESTION	PLANNING NOTES
Centering Cultural and Community Funds of Knowledge	*How does my lesson help students connect mathematics with relevant/authentic issues or situations in their lives?*	• Planning of tasks related to students' lives, lived experiences, and interests • Planning of tasks related to school experiences that all students could connect to • Positioning students as doers of mathematics and experts of their invented strategies through allowing time to share strategies with partners • Recognizing and drawing on the funds of knowledge that each individual student brings to the mathematics classroom • Allowing students to defend and justify their mathematically brilliant strategies (not pigeonholing students into teacher expected strategies)
RIGOR & SUPPORT DIMENSION	**ESSENTIAL QUESTION**	**PLANNING NOTES**
Affirming Multilingualism	*How does my lesson make space for multilingual learners (MLL) to be central participants in mathematics activities?*	• Purposeful partnerships to share strategies based on home language buddy • Language supports to ensure all students have access, such as sentence stems, visuals, videos, and translating tasks into different languages • Highlighting MLL students as experts within the mathematics community • Opportunities to share insights publicly in home language with a language buddy • Creating effective co-teaching structures that include identifying the roles of MLL teachers in the math classroom

Mathematize the World Through Images

We were eager to implement the strategies and activities discussed to ensure that we were supporting all students in affirming their math identity and creating tasks that were relevant and interesting to students' lives. We decided to focus on helping students mathematize their world through the use of images, videos, and relevant tasks, specifically designed to give access to all students, including our multilingual learners. We worked together to design high-level cognitive demand tasks that connected to students' lived experiences, school experiences, cultural backgrounds, and interests. We wanted to ensure that, each day, our task engaged students in our classroom and helped them to connect with and relate to mathematics.

We also wanted to ensure that we were creating and implementing tasks that gave space to multilingual learners to be central participants in mathematics activities. We were very purposeful in planning how we could support students in using mathematical language as well as their home linguistic repertoires in translanguaging.

For example, Figure 7.6 showcases a few components of one day's mathematical lesson that supported the state standards of interpreting multiplication and division word problems using a task about fish in a pet store. Each component of the task is specifically designed to elicit language use, **Affirm Multilingualism**, draw on students' **Funds of Knowledge**, and support students in their use of understanding and using mathematical language.

Zoom in: Launching the Task

The task in Figure 7.6 is presented to students. The fourth graders all start smiling and yelling "Faraz!" when they see the first slide with images of betta fish. Faraz is a neurodiverse, multilingual student. The children in this classroom know that Faraz has a passion for learning about and sharing his knowledge of fish and that he regularly visits local pet stores with his family to look at all of the interesting fish. He also has betta fish at home that he talks about often.

The class is immediately engaged in the story problem, eager to share their ideas after looking at the pictures and watching a video about betta fish (and gaining some new fish facts from Faraz in the process!). Students are sharing their thoughts about what they might be solving today. The use of language is evident everywhere, as students are excitedly sharing their ideas about what they notice and wonder. As students go off to solve the problem independently, we notice that the children are happy to see others so enthusiastic about the math task for the day. Students choose a strategy to solve the problem, while we walk around the room, asking purposeful questions and making note of student ideas. Students are purposefully paired, based on

FIGURE 7.6 Task Components

COMPONENT OF THE TASK	PURPOSE
Launching the Task Betta Fish Tanks Task	Students view a picture and/or video related to the context of the task. Students are asked, "What do you notice? What do you wonder?" Contexts are carefully chosen, based on students' school experiences, lived or cultural experiences, or student interests.
Understanding the Task—Ensuring Access for All Students **What do you notice and wonder?** New Betta Fish have just arrived to the local pet store! The pet store only allows 4 betta fish to live in each tank. If there are 92 new betta fish, how many tanks does the pet store need for the fish? What do you notice? / What do you wonder?	The task is presented, first as a numberless word problem so students can focus on the context and language of the task, before jumping to the mathematical operations. Students are given opportunities to notice and wonder individually, then by turning and talking to a partner (could be a language buddy), and then sharing with the class. All student thinking is honored and valued as some students share ideas related to the mathematics, others share ideas related to the story, but all ideas are recorded and celebrated. This is also the space where students extend their connections from the initial image of fish.
Independent Work Time & Purposeful Partner Sharing New Betta Fish have just arrived to the local pet store! The pet store only allows 4 betta fish to live in each tank. If there are 92 new betta fish, how many tanks does the pet store need for the fish? Label the quotient, divisor, dividend in your work. I found the quotient by _____. I made _____ groups of _____. I split _____ into _____ equal groups. I divided _____ into _____ equal groups of _____ ¡Nuevos peces betta acaban de llegar a la tienda de mascotas local! La tienda de mascotas solo permite que Vivian 4 peces betta en cada tanque. Si hay 92 peces betta nuevos, ¿cuántos tanques necesita la tienda de mascotas pra los peces? Etiqueta el cociente, divisor, dividendo en tu trabajo. Encontré el conciento por_____. Hice_____ grupos de _____. Divido _____ en_____ grupos iguales. Dividí _____ en _____ grupos iguales de_____	Students have time to work independently and try any strategy of their choice. The teacher purposefully pairs students based on a number of factors, including goal of the lesson, strategies, languages, and so on. When students share with their partner, they use the sentence stems to guide their conversation and practice using mathematical language. Students can choose to work on this or record their thinking in their home language, and language prompts are given if students choose to use them.

SOURCE: 2 betta fishes in 2 tanks image by aluxum/istock.com; pet shop image by niunium/istock.com; betta fish images by anurakpong/istock.com, CmErsoy/istock.com, NatalyaAksenova/istock.com *strategies and language support, and students are thrilled to be sharing their work with a partner. It is a noisy classroom—one that is full of productive and engaging mathematical discussions. Faraz, a student who historically may have not felt welcomed to share his mathematical ideas, is committed to solving the problem and excited to share his strategy.*

This example is one small sample of what we noticed in our classroom on a daily basis as we focused on *(Re)Humanizing Mathematics* through building on students' *Funds of Knowledge* and *Affirming Multilingualism*. As we continued this work, we noticed a drastic shift in *who* was doing the talking in the classroom. Our focus on creating tasks relevant to our students' lives was making a huge impact on the level of student discourse and giving our MLLs an abundance of opportunities to communicate their thinking and reasoning, as well as creating space for all students to be central participants in mathematics activities.

Affirming Positive Math Identities for All Students

As we continued the work of *(Re)Humanizing* the mathematics block during our fourth year in our coaching partnership, we also wondered: *How might these changes affect the mindset of our students? How do our students feel about math? How might our students' feelings change throughout our journey?*

We decided to conduct student interviews to track how the change in the structures of the mathematics block and the focus on *(Re)Humanizing Mathematics* could impact the students experiencing this shift. We conducted interviews at both the beginning and end of the year, and we were interested to learn how this radical shift in our practice might change students' perspective around mathematics. During the interview, students were asked questions, such as the following:

- Tell me about what conversations in math sound like.

- How do you feel when you share an idea during math?

- How do you know when someone is listening to your ideas?

- How do you show others that you are listening to their ideas?

Figure 7.7 highlights the drastic differences in our students' responses to the same interview questions, when asked before and after the focus on *(Re)Humanizing Mathematics* was implemented in this specific classroom. The "before" column shows a child's response at the beginning of the year, and the "after" column shows how that same student answered at the end of the year. The change in student responses demonstrates the large impact that the CRMT2 framework had on our students' beliefs about seeing themselves as mathematicians.

FIGURE 7.7 Before and After Interview Responses

BEFORE	AFTER
When asked: How does sharing an idea in math make you feel?	
They aren't talking, I'm not talking	Happy because I am giving someone else a chance to understand math.
Kind of nervous and like I'm going to make a whole bunch of mistakes. I always get scared and make excuses.	**I'm not afraid, I'm brave because I want to represent to my classmates in what they don't know, and I can teach them.**
When asked: What do conversations sound like in math?	
Talking about math and not about anything else (like what you brought for lunch)	**What their strategy is and how they did the problem and like sharing what you did and then discussing what strategy to use and discussing how you can make it better.**
I don't know. I'm not that good with math. In first grade I was really good. I always got like 150 and 165. Now it's even harder and I go to a lot of math classes, and it still doesn't work.	**Saying mathematical words and agreeing and disagreeing and making a picture in your mind.**
When asked: How does math make you feel?	
A little scared	Math makes me feel happy because I get to learn new strategies and work with different people to learn their strategies, too.
Nervous	The way I feel about math is that it is really fun because I think multiplication and division are getting easier because of the partners and the strategies, and it's really nice to make a picture and have partners to understand you.

We sit down together for a coaching conversation following the second round of student interviews. We are curious to reflect on how student responses have changed and the impact that this new learning structure and focus on **(Re)Humanizing Mathematics** *has had on the students in their mathematics community. We are astonished to reflect on the responses and learn just how much these changes in practice using the CRMT2 framework transformed students' mathematical journeys and their feelings toward math and math learning. During this reflection, one of the major takeaways that our team notices is that math in this particular classroom went from giving students feelings of nervousness and fear to feelings of courage and happiness. This is a proud moment for us, as we contemplate how our teaching evolution has affirmed the positive math identities for all students who we worked with in these classrooms.*

At this point in our journey, we had implemented a structure that **Disrupted Status and Power** through the drastic change in the design of our mathematics block and through our deliberate efforts to position students as doers of mathematics. This next step focused on **(Re)Humanizing Mathematics** by connecting to students' fund of knowledge and honoring the diversity and expertise of MLLs in the classroom; but there was still more to come in our journey to use the CRMT2 framework to become more culturally responsive educators.

YEAR 5: A FOCUS ON ANALYZING AND TAKING ACTION

During the last year of our coaching partnership, we recognized what was blatantly missing from our ever-evolving CRMT: opening mathematics up as a space for students to **Analyze and Take Action**. In a way, this felt like the biggest shift yet: How do we connect mathematics to this larger critique of injustice? One thing that we recognized was that, while *every* lesson might not be one focused on social justice, it was imperative that we design *some* lessons for students to have this opportunity. Using the CRMT2, we began to explore what this might look like in our fourth-grade classroom.

We started with a previous task that Holly had worked on in some other classrooms (Tate et al., 2022). The Library task is one in which students consider whether or not the racial and cultural representation of the main characters of books in their library is fair:

LIBRARY TASK

How fair is our library book collection? Create a plan to help figure out if our library has enough books with main characters from different racial and cultural backgrounds.

However, for this iteration, we wanted to really expand on children's awareness of their impact on social change. We wanted the critique to connect to their libraries but also for students to understand the role of race and culture in their lives. We wanted the action piece to include the space for students to have a platform to use their voices for change while also having firsthand experience, so we spent a large amount of time planning this process out using the CRMT2. Our planning is captured in Figure 7.8.

As expected, students found that there were a substantially larger amount of white and animal main characters than characters of color, as shown by the graph in Figure 7.9. A student group comments on the unfairness of their

FIGURE 7.8 Reflection on Learning Within the CRMT2, Focus 3

POWER & PARTICIPATION DIMENSION	ESSENTIAL QUESTION	PLANNING NOTES

Analyzing and Taking Action	*How does my lesson support student use of mathematics to analyze, critique, and address power relationships and injustice in their lives?*	• Task: Is our library fair?
		• Prework around race and culture: culture map, reading a book about race and our skin, morning meeting conversations and reflection about why we are special, looking at national data of race of main characters, looking at district data of what races make up our community
		• Analysis:
		  ○ Students create a plan to make our library more fair
		  ○ Students choose which kind of graph to create to show others their findings
		  ○ Students write up critiques of what this means that they can present to others
		• Action:
		  ○ Students work within a budget to purchase new books for their classroom library, as well as every other fourth- and fifth-grade classroom.
		  ○ Students create a new graph to show how their purchases changed representation and create a new analysis
		  ○ Students will unbox, label, and deliver books to classrooms
		  ○ Students will create a slide deck legacy presentation to share their findings with classes and others, including why they chose the books that they did and what else remains to be done

classroom library after analyzing a sample, describing this analysis as part of their legacy presentation to others. They write, "This is our chart down there it shows how many books each group got. We made it after we sorted a couple of books. It showed us it was very unfair that there are nineteen white books and only three Latinx books. So we decided to solve the problem by taking action."

With the focus on critiquing the world, however, we were amazed at how our students interpreted and made sense of these data and how they spoke to the inequitable systems at play in the publishing of children's books. One child wrote,

> *"The library needs different race books because it can't be the same thing like animals and white so it can change. So african [sic] and other races. I think book companies need to know this because they need to know how many books they make of every different kind of books."*

FIGURE 7.9 Example of Student Graph

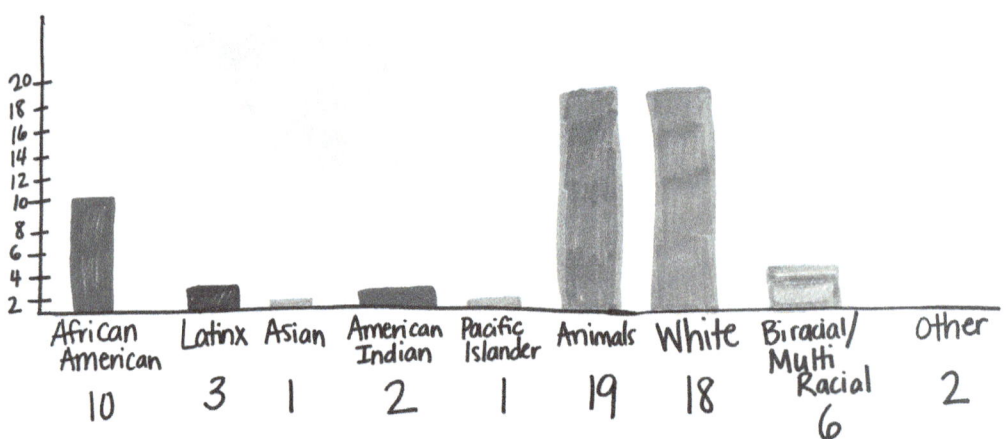

Further, we noticed how the task elicited feelings of empathy and compassion. Many groups began their presentations by framing how this work made them feel. One group wrote,

> "We did not see ourself in books. So we felt upset and not good so we said this is unfair. Pacific islander had no books so we felt very disappointed and sad that some races had no books about there self."

Another group wrote about their plan to purchase books, noting the need for others to see themselves in books. Specifically, those children who had zero main characters who reflected their race.

> "In order to make our library more fair we used a budget of $50.00 to make our library more diverse so that people can see themselves.
>
> – Bought more diverse books
> – Made our library more equal
> – We knew it was unfair so we bought more books of race
> – We did this for people to see themselves in other books
>
> We used our graph to make decisions about what type of books to buy. We bought diverse books. So the ones that have less representation or 0 books we bought them."

Children began to see that, while we may have created some change, there was still a lot of work to be done, seen in Figure 7.10.

Ultimately, this month-long endeavor ended up being one of the most rewarding experiences for both the students and for us as mathematics educators. Ten-year-olds realized that not only can mathematics help them to

FIGURE 7.10 Examples of Student Slides During
the Legacy Presentations

Now Our Graph Looks Like This...

We made our classroom library fair but the books are still not completely fair. We do have more diverse books .

- Future students could talk to teachers, principals, adults, and students to diversify other libraries
- Scholastic should make more diverse books

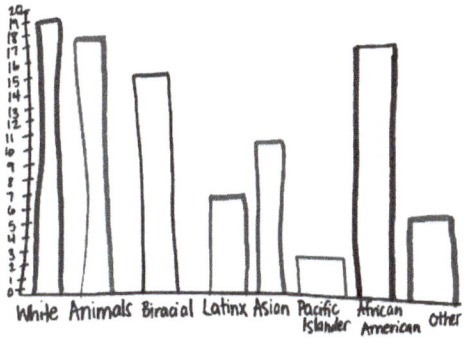

Others Need to Know

we were able to make our classroom library more diverse. it is important for other people to know that we helped our library and other classrooms more diverse .It is helping make libraries and classroom libraries more fair, but there's still more to be done.

- Others need to know..
- We wanted to fix the classes to see are self in book
- Other people don't see thar self in books

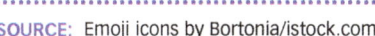

SOURCE: Emoji icons by Bortonia/istock.com

critique the world, it can also be a space where they can take action, and they felt so proud to be a part of that change.

> "At the end of the project we felt helpful because we helped many people see themselves in the books. OUR group felt really proud and helpful we are amazed with the work we did!"

LONG-TERM LEARNING FOR THE CENTERING OF CHILDREN AND THEIR IDENTITIES

This chapter describes a five-year partnership between an instructional coach and teaching partner using the CRMT2 framework to guide their journey of becoming more culturally responsive in their mathematics teaching practices. This continues to be an ongoing and flexible journey. Kaitlin and Holly took a unique pathway, learning about and focusing on a few different components each year. The CRMT2 acted not only as a lesson analysis tool but also as a tool to help change the fundamental structures of our mathematics classrooms. The dynamic ways in which the dimensions weave together allow for teachers, teams of teachers, or schools to have creative and personalized paths to becoming culturally responsive teachers of mathematics in ways that make sense to them. As they began to understand the ways that these strands interweave with one another, they recognized that there is power behind the sets of three dimensions within the strands. Any direction taken to make sense of the culturally responsive matrix was met with a dimension that was unique to ambitious mathematics teaching and not a part of previous training around math reform-oriented practices such as *Disrupting Status and Power*, *Centering Cultural and Community Funds of Knowledge*, and *Analyzing and Taking Action*. As teachers embark in their journeys to become culturally responsive mathematics educators, we urge them to be critical and push their thinking beyond what might be comfortable toward continued growth and learning in an effort to be better for each and every student as we transform the math classroom.

DISCUSSION QUESTIONS

- Reflect on the CRMT framework discussed in this chapter. Which dimensions are already present in your classroom? Which dimensions can you commit to learning more deeply about in your classroom immediately after reading this chapter?

- Think about the student voices in your classroom. Whose voices are heard? Whose voices are not heard? How can you implement the dimensions of the CRMT1 framework in your own classroom to ensure visibility of *all* students?

- *Disrupting Status and Power* is one focus of this chapter. Reflect on the students in your classroom and the stereotypes or beliefs that might surround these groups of students, either directly within your setting or within the larger system. What are these stereotypes or beliefs? How could you use the CRMT2 framework to work against them?

CHAPTER 8

....

CORAZONES MATEMÁTICAS: CRMT IN BILINGUAL CLASSROOMS

By Melissa Adams Corral

In this chapter, a bilingual Latina educator working in an urban setting, Melissa Adams Corral, shares an account from her teaching that can help unpack the way the Culturally Responsive Mathematics Teaching Tool (CRMT2) dimensions can unfold across one lesson. This problem-solving lesson takes place in a second-grade, dual-language classroom, where the teacher had developed a practice that centered bilingual students' voices and thinking. Let's hear from Melissa directly, as she explains more about how centering her students and teaching in alignment with the CRMT2 dimensions allowed her and her students to feel more welcome in the world of mathematics.

BECOMING A BILINGUAL MATHEMATICS TEACHER

If you had told me (Melissa) when I first started teaching that I would come to love mathematics, I would never have believed you. Las matemáticas had never held sense or meaning for me. Yet, during my years in the classroom, listening to children's mathematical thinking changed everything—they taught me what it really meant to center students' voices and thinking but also to find the joy and wonder in mathematics. Every day brought excitement—I could not wait to see how they would approach the tasks that I designed. As I planned cross-curricular units, I would challenge myself to find what could be mathematized and realized how much math had become a tool that I used to make sense of the world. You see, when you make your math class a space of exploration and discovery, it isn't just the children who benefit—you grow, too. My journey to re-envisioning what a mathematics classroom could be began with the heart—with the impulse that propelled me into teaching in the first place.

I went into this profession, first and foremost, because of the possibility that came with the title of bilingual teacher, a role that did not exist when I was in school. I wanted the children I shared a classroom with to be fully proud of themselves, of their language and their culture, in the way that I had become proud of these things as an adult. I could remember being a young girl and feeling like those parts of me were never known or recognized by my teachers—they never asked about what I learned about the world when I spent summers in Tegucigalpa, Honduras, nor did I ever have the opportunities to write in español in any classes other than my hour a week of Heritage Language. My all-English math classes never knew that I was memorizing my tablas de multiplicación con mi mamá, they only told me that I needed to get faster, not worry about whys or hows. Over the course of my years in the classroom, I tried to make it clear to my students that I saw them, que los adoraba tal y como eran, that they were the reason for everything we did all day and that what we did should make them feel seen, heard, and reconocidos. So, when rules and policies limited space for my students to talk, to be themselves, to take up space in the classroom with their bodies, their voices, their ideas, well, those rules just had to go.

The example lesson in this chapter comes from one day in my second grade, two-way dual language classroom in central Texas. At the time, our district had a policy requiring strict language allocations and separations for all content areas. Mathematics, according to this policy, needed to be taught in English. This decision was justified with the profoundly misguided assertion that "math is a universal language," as if in a math classroom all we do is produce symbols that would be recognized any and everywhere. What that meant for my second graders was that, when the school year started, my Spanish-speaking students were used to being quiet until math was over. This was not the first time I had seen this, nor was it the first time that I had broken this rule. In previous years, I had found that my bilingual students had been silenced in their all-English math classes and that ensuring they had access to and the right to use of all their language practices had helped to build positive math identities (Adams, 2018).

As I started teaching my second graders, I knew that they had received math instruction solely in English during kindergarten and first grade. I also knew that I wanted to design a mathematics classroom where all students' ideas were recognized and celebrated. To do that, I explicitly designated math as a bilingual instructional time where all members of the classroom community were welcome to fully exercise our language practices by translanguaging (Maldonado et al., 2020). When I say translanguaging, I am referring to the right to flexibly move across the social boundaries of language—those can be the perceived boundaries of named languages such as English or Spanish (Garcia et al., 2017) but also the boundaries that tend to limit when and how children are welcomed to speak in classrooms. By the end of the year, we had really come to build our entire mathematics block as a designated translanguaging space. Doing so allowed us to address the three main strands of CRMT2: Knowledges and Identities, Rigor and Support, and Power and Participation (see Figure 8.1). But it might be helpful to imagine what that looked like within one simple daily lesson routine.

FIGURE 8.1 CRMT2 Overview

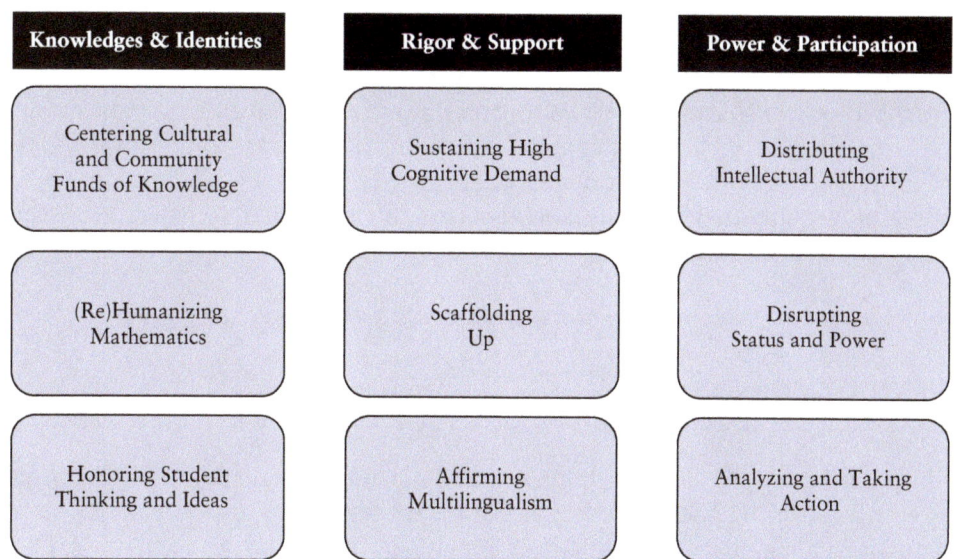

Knowledges & Identities	Rigor & Support	Power & Participation
Centering Cultural and Community Funds of Knowledge	Sustaining High Cognitive Demand	Distributing Intellectual Authority
(Re)Humanizing Mathematics	Scaffolding Up	Disrupting Status and Power
Honoring Student Thinking and Ideas	Affirming Multilingualism	Analyzing and Taking Action

THE POLLITO PROBLEM

My students and I engaged in problem-solving every day, based in large part on the framework of Cognitively Guided Instruction (Carpenter et al., 2014). While many of the problems that we posed were, to a certain degree, contrived, I tried to connect our problem-solving lessons to what we learned in cross-curricular units (like, for example, one unit on the water crisis in Flint, Michigan, and Crystal City, Texas) or to my students' lives. There was one week where students were amused to find that each day there was a new problem featuring Cody and his younger cousin. Cody was a quiet and calm boy, but his cousin was known for being rambunctious and constantly getting him in trouble at home. We did a series of problems where Cody's cousin took things from his collections or tried to get a greater share of a snack than him. As the week went on, the kids were excited to try and guess what she would do next. When she heard she starred in a week's worth of problems, she was delighted to be the center of her cousin's class's attention.

One day, during the morning meeting, Kelis told a story about having gone with her family to visit the Pulga Ocho Doce that weekend and how the highlight of the trip was that her parents bought her a pollito. All the students were excited to hear about her new pet and asked her to bring it to school. Kelis asked whether she might be allowed to bring her pollito the next day, and her mother and I coordinated a plan to have the baby chick visit the classroom. When the pollito arrived, the students sat in a wide circle on the edge of the carpet and watched it peck around in the middle. We fed it small pieces of bread that Kelis's mom had brought, and Kelis showed us how she held the chick.

That weekend, while lesson planning, I knew I wanted my students to continue to work on groups of ten problems to strengthen their place value understanding. They had been struggling to subtract once our daily number talks went from equations like 65 – 12 to ones like 62 – 15, and their typical strategy of tens minus tens and ones minus ones was proving to be complicated (eventually some students discovered negatives and stuck with the method; others found new ways to break up the minuend). I thought about the importance of reinforcing their place value understanding and decided to pose a groups-of-ten problem (Carpenter et al., 2014). When I tried to imagine a context, I remembered the bag of bread that Kelis had fed the baby chick from and wrote the following problem:

POLLITO TASK

Un pollito come 10 trocitos de pan cada día. Una bolsa de pan trae 117 trocitos. ¿Cuántos días podrá comer el pollito de esa bolsa?

A chick eats 10 pieces of bread each day. A bag of bread has 117 pieces. How many days will the chick be able to eat from that bag?

In our classroom, problem-solving was all about students expressing their thinking—to me, to and with one another, and even to themselves. A major goal of our classroom was the recognition that everyone had mathematical ideas that mattered, that there were many approaches to thinking about the same problem, and that who we were was a part of doing math—from the things we knew and cared about, to the languages and forms of expression we used when sharing our thinking. With this problem, I knew that Kelis would feel excited to see her pollito brought up again. I also knew that everyone in the room had shared the experience of seeing Kelis's pollito from the Pulga Ocho Doce pecking at our classroom carpet. They were also very familiar with our problem-solving routine and with being the center of what it meant to do mathematics.

Launching the Task

The day we solved the pollito problem started like every other day. We held our morning meeting where we shared something about our morning using the language of the day. Then, I collected any coins that students had brought in, and we identified their value and added them to the jar of change we were collecting for Flint, Michigan. Students had wanted to contribute to supporting the community whose struggles they had learned about and would bring in spare change whenever they could—that day we added in coins that took our total from $36.88 to $38.59. This action project had also become a daily routine that emphasized certain aspects of numeracy with coins, introduced the use of the decimal point authentically, and gave purpose to practicing skills.

After counting our coins, I had students take out their problem-solving notebooks and get their problem slips from me. They could choose to glue in a problem in English, Spanish, or both. This was typical, and as we launched the problem, I paid careful attention to ensure that we broke down the big ideas in both English and Spanish, to ensure that everyone in the room had linguistic access to the task. To start, we read the problem in English, and then I asked for a volunteer to read the problem in Spanish. I was surprised when Arturo raised his hand. He was an expressive and confident reader and writer but tended to feel much more insecure when it came to working with numbers. In math class, he often sought me out for assistance while I was circulating in the class observing the different strategies being used. One goal I had for Arturo was to seek support from peers when he felt challenged and also to give himself time and trust to figure things out. This was explicitly connected to my efforts to ensure that both students and I were seeing intellectual authority distributed across the class community. As Arturo read in Spanish, he stopped at the number 117 and looked up at me: "Qué es ese número? (What is that number?)" I replied, "Tú tienes que decírmelo (You tell me)." There was a long pause as Arturo thought deeply about what to call this three-digit number. Recognizing that he could use support, I asked the class if "¿Alguién puede ayudar a Arturo? ¿Cómo se lee ese número? Can anyone help Arturo? How do we read that number?" and then encouraged Arturo to choose a classmate whose hand was raised. This was intentional—I wanted Arturo to recognize that his peers were a resource he could call upon for support and that he did not have to only rely on me in moments of struggle. He chose a peer whose hand was not raised— Anthony. Anthony, whose parents were bilingual, felt much more comfortable speaking in English and immediately shook his head no. I leaned over and suggested that Arturo ask Emanuel, whose hand was raised, to jump in. Arturo called on Emanuel, who supplied "Ciento diecisiete." Arturo repeated the number and finished reading the problem.

After hearing the problem in English, I asked my class our usual questions: "What do we know?" and "What are we trying to find out?" As students answered these questions in Spanish, I scribed the ideas in English (see Figure 8.2), again ensuring access to the conversation across the linguistic range in our classroom.

At this point in the lesson, I wanted to make sure the whole class had the necessary access to the problem, allowing their focus to be on their chosen problem-solving strategies. Those strategies, which they would work on

FIGURE 8.2 Problem Launch

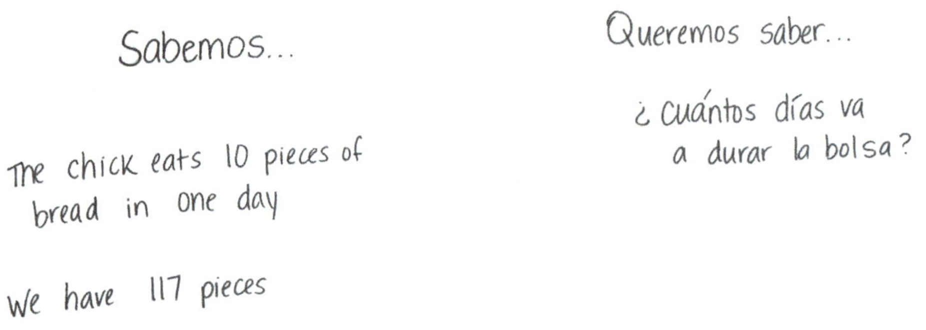

Sabemos...

The chick eats 10 pieces of
bread in one day

We have 117 pieces

Queremos saber...

¿ Cuántos días va
a durar la bolsa?

where and with whom they chose, would be the focus of our discussion and, ultimately, the lesson. As our launch was wrapping up, Aliyah spoke up to answer the question of what we were trying to find out before it had been posed. As soon as we had charted all the information we knew, she offered: "So, basically we're trying to figure out how long the bag's gonna last." This contribution was revoiced and then written in Spanish on our anchor chart. With that, the class was off to solve in the ways that they chose.

While circulating, I started out by talking to AJ, who had called out the answer as we left the rug. He had a fantastic mental strategy, explaining that the bag would last for "11 days and 7 hours," because "10 days is 100, 11 days is 110, and 12 days is 120." I encouraged him to find a way to show his thinking on paper. While AJ loved solving problems mentally and explaining his thinking, he often became frustrated with finding ways to represent his ideas in writing. He seemed particularly excited about his thinking that day and eagerly grabbed his pencil. Eventually, he produced the representation of his thinking about the problem in Figure 8.3.

Trabajando Together

As I continued circulating, I noticed Maricela and Chantel working together on a calendar of days, with the number 10 written in each day in Chantel's calendar and Maricela's days each containing ten dots. Chantel was new to our class. She had been moved to our room from an all-English classroom due to conflicts between her and her previous classroom teacher. Maricela was typically very shy—she had come from a different school, and her mother had asked for us to support her in seeking special education services. I was called over to the carpet by Gabriel, a confident bilingual student who usually worked on problems with his friends. Today, he was having trouble figuring out where to start. We walked back to the carpet together, where many students were sitting and working, including Arturo. I sat cross-legged next to Gabriel, and as soon as I had asked him if he remembered seeing the pollito when he visited our classroom, Arturo had turned his body toward me and scooted much closer. In the same way that I hoped Arturo might look

FIGURE 8.3 AJ's Written Representation of His Thinking

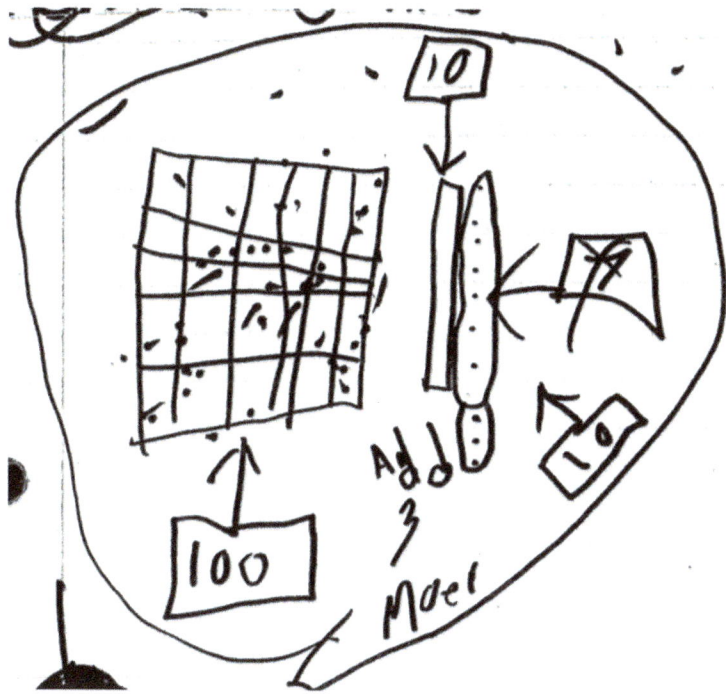

to his peers as resources for solving problems, I also knew that many of his peers had a negative perception of him. He had, over the course of their years together as a class, come to be seen as someone who had less mathematical knowledge to offer. I was often thinking about how to ensure that all students, but especially those perceived as less mathematically capable by their peers, were featured as problem-solvers.

I asked Gabriel how much the pollito ate en un día, and he answered "Diéz." Then I asked how much he ate en dos días. Meanwhile, Arturo had scooted as close to us as possible. His notebook, which he was continuing to work in, lay in front of him and in between him and us. He was drawing out rows of ten circles and carefully writing the count next to each row. At this point, he had gotten to 60 pieces of bread. Gabriel, after a couple of seconds considering the question of how much the pollito would eat in two days, answered, "Veinte," and I followed up with, "Y en tres días?" He knew that was thirty, so I asked how many pieces I had to give to the chick. He answered, while, in the background, students had lively discussions about the numbers in Spanish. Gabriel and I then discussed what we were trying to find out—how long the bag of food would last—and whether there was enough for three days. His friend, Juan Carlos, had joined us by that point and answered that there was enough for three days. At that point, I asked Gabriel and Juan Carlos if they could think of ways to keep feeding the chick after the third day. While I waited for them to think,

FIGURE 8.4 Arturo's Strategy

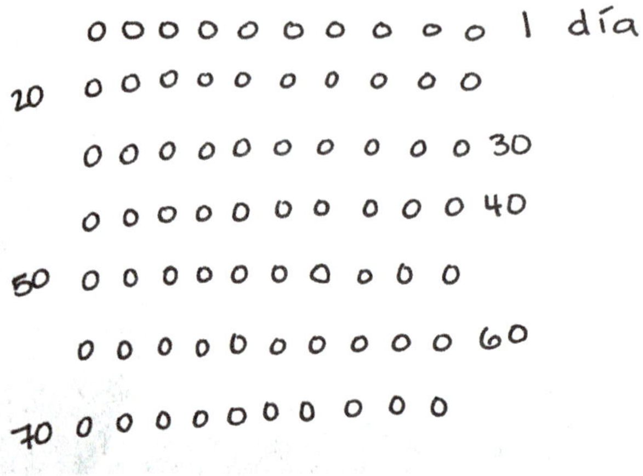

ooooooooooo I día
20 oooooooooo
oooooooooo 30
oooooooooo 40
50 oooooooooo
oooooooooo 60
70 oooooooooo

Arturo jumped in, pointing at his work and announcing, "Son sesenta (There are 60)."

I asked how many days it was, while turning Arturo's notebook (Figure 8.4) so Gabriel and Juan Carlos could look at it—"Mira lo que está haciendo Arturo." I then showed them how he had organized ten for each day and asked them how many days he had. Arturo jumped in again: "Vamos a llegar a este número [pointing at the 117 in the problem]." You would not know, at this moment, that Arturo was anything other than a confident problem-solver. "Does that make sense? ¿Qué notas que está haciendo, Arturo?" I asked the other two boys. "He's counting every day," said Gabriel. As we analyzed Arturo's work, I asked how many days he had on his paper again, and this time Arturo answered, "Siete."

From there, Arturo continued to work while collaborating with Gabriel and Juan Carlos. When he reached 100, he became concerned—how could he get to that number, 117? Juan Carlos, who at this point of watching Arturo's strategy felt more sure of his thinking, said, "Well, you're at 100. That means you just need 17 more." Gabriel agreed, "Solo tienes que poner 17 más." Arturo asked "¿Cuántos dieces hay en diecisiete?" He was organizing his rows by tens, so he wanted to know how many more rows of ten he would need to draw in order to get to seventeen. Once he had drawn ten, Juan Carlos told him all he needed was seven more. Gabriel and Juan Carlos then confidently began to write in their notebooks—having watched Arturo make the features of the problem visible to them, while also being able to contribute to his work, they now had a sense of how to document their own thinking.

FIGURE 8.5 José's Strategy

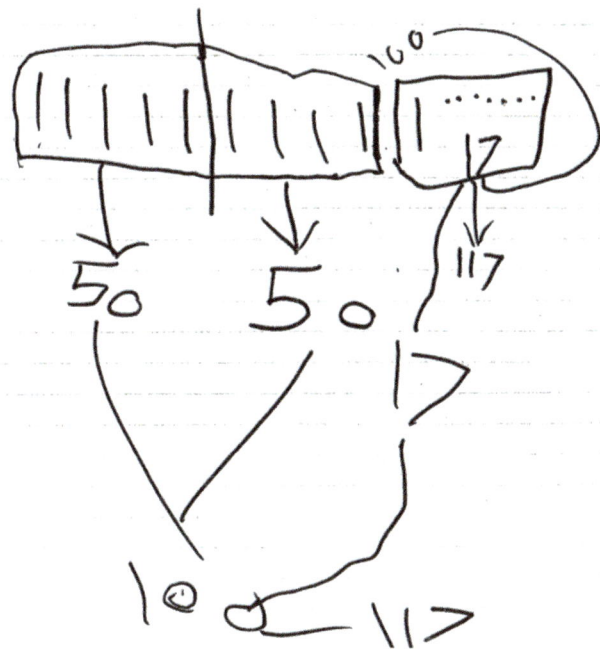

As I moved to start calling students to the carpet, I saw that one student, José, had become upset and crawled under a table. Earlier, before working with Gabriel, Arturo, and Juan Carlos, I had checked in on José. He had a wonderful strategy for visualizing the number 117 (see Figure 8.5).

When I approached him, I realized that he was upset because he "didn't get the answer." While I had spent a lot of time during the year moving the class away from an emphasis on answer-getting and toward a focus on students' strategies, many of them still carried the idea that math was only about finding right answers. It can be hard, when your math instruction involves a shift away from students' previous experiences, to convince them that you really aren't focused on seeing whether your students are right or wrong. I looked at José's work again, which he had left on top of his table and thought that, perhaps, we could use the visuals of his strategy as a class and show him that he had done valuable mathematics work, answer or no answer.

Discussing Our Strategies

I gathered my students back together at the carpet and told them that today, we would be looking at José's strategy because I thought that they might like it. Back on our classroom notes page, I drew out ten lines, like in his notebook, and circled them (see Figure 8.6). Then I drew an additional line and seven dots to the right of that, and asked the class, "What did José do?"

FIGURE 8.6 Evolving Chart: José's Strategy Is Shown in Lines and Dots

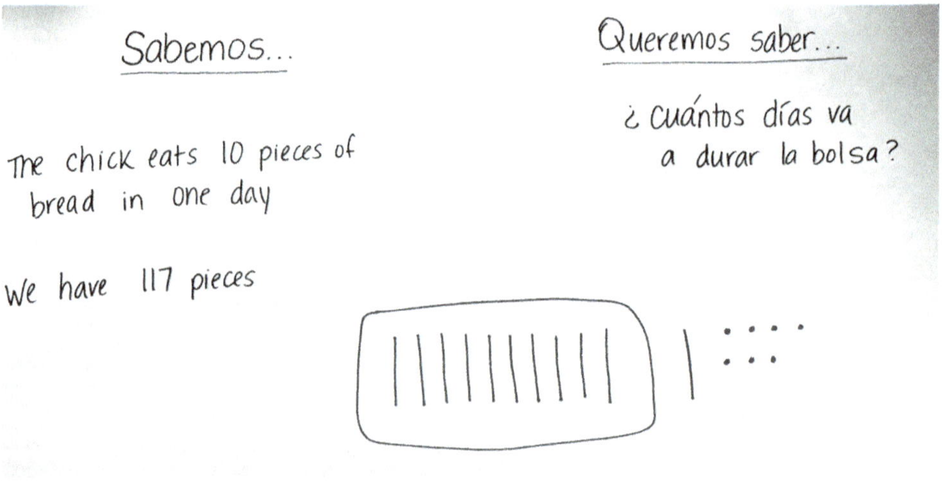

Sabemos...

The chick eats 10 pieces of bread in one day

We have 117 pieces

Queremos saber...

¿Cuántos días va a durar la bolsa?

There was a burst of voices and ideas—many students could see the sense in this drawing.

I asked Aliyah to share her thinking, and she said that José "drew the picture of the number." I revoiced her idea and asked, "Of what number?" Genesis and Kelis both answered, "One hundred and seventeen." I asked the class how we knew that was a picture of 117. Aliyah and Kelis each began explaining their thinking—Aliyah in English and Kelis in Spanish.

> *Aliyah:* *Because … [begins pointing and mouthing a count]*
>
> *Kelis:* *Porque aquí [pointing at the encircled ten lines with her pencil] hay 100, y aquí [pointing at the one line] hay 10, y aquí [pointing at the dots] hay siete. Because here there are 100, here there are 10, and here there are 7.*
>
> *Aliyah:* *[picking up after finishing her count] There's ten tens, which, that makes one hundred. Then there's one ten that makes ten, and then seven ones.*

I affirmed both of their ideas. I loved hearing the same idea in both languages—it seemed as if the students themselves were ensuring the balance and access that I had sought when establishing our math classroom as a translanguaging space. I had not forgotten about José, however. I asked the class why he would draw that picture to solve the problem. I called on Emanuel, and Vanessa jumped up and said, "The tens are like the days." Emanuel added, "Because he's counting by tens." To ensure we heard both of their ideas, I repeated Emanuel's idea and asked Vanessa to restate her point, then turned to the class: "How many of you put ten for a day?" Many hands were raised. I pointed out what I had seen while circulating, "Arturo drew each one, I saw Maricela put ten dots in a calendar, so, José, this is kind of like what they did." Together, we counted the days, labeling them with real days of the week like Maricela and Chantel had. Chantel was excited to share that "the only way we did it, like, the same is because we helped each other."

This was a powerful statement from Chantel—one that worked to counter the ways both girls were seen within the classroom community. Chantel was perceived as someone who could be mean, in part because of her experiences in other classrooms. Maricela was seen as needing academic help, not as someone who helped her classmates. In Chantel's words, however, she rewrites both of those identities, making herself into someone who is helpful and kind to her classmates and also asserting that she also received help from Maricela. The space given to students to choose their own strategies, workspaces, and partners in many ways offered the opportunity to enact new stories for themselves and for their peers and to share those stories so that they might add to the community's understanding of who they were.

As I continued labeling the days of the week on José's tens, Arturo piped up without being called on—something that was not typical of our problem-solving discussions. Of course, he had been in the role of teacher already during this lesson. "Como es un número grande, va a tener once días el pollito para comer. (Since it's a big number, there will be eleven days for the chick to eat.)" This was an opportunity I would not miss—I revoiced Arturo's ideas in English to make sure the whole class was aware of his contribution and asked the class if they could see eleven tens. Vanessa declared, "You can see it in the number 117." When asked to elaborate, she jumped up and used two fingers to point to where she saw 11 tens on the chart (see Figure 8.6).

As our discussion began to wrap up, I addressed the factor in this problem that had most troubled many students in the class. "What was the part that was the most 'ugh' in this problem?" There was an immediate chorus of responses saying "seven." As I had circulated, many had noticed that it was not an even division. They were so bothered by the seven. Yet, as they asked about it, I challenged them to think about what would happen in real life if there was not enough food left to feed the pollito that day. Genesis's work (see Figure 8.7) shows this idea really was important to her thinking about the problem.

Her paper reads, "If you give it those seven pieces, you can go to the store and buy more food." For students like Genesis, being able to imagine the problem in the real world was an important component of problem-solving.

FIGURE 8.7 Genesis's Strategy

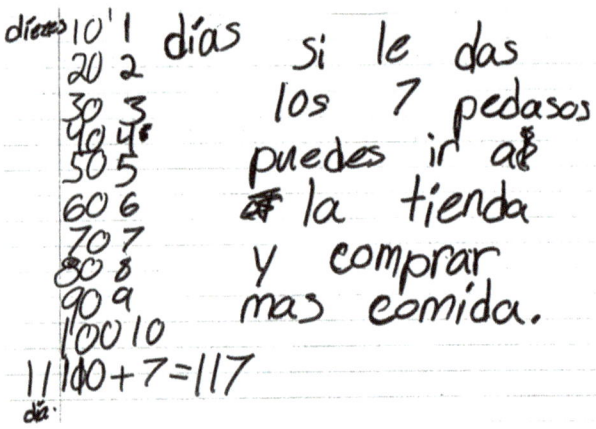

Using the CRMT Lesson Analysis Tool

When I go back and watch video of this day again, I am, of course, watching with years of experience that I did not have on that teaching day. As I go through the elements of the CRMT Lesson Analysis Tool, there are aspects of my teaching that day that I can challenge and imagine improving. At the same time, I am also able to recognize the many aspects of what happened on that one day in our second-grade year that were doing much to ensure a culturally responsive experience with mathematics.

PAUSE AND REFLECT

- Were there aspects of the CRMT Lesson Analysis Tool that you identified as you read the story of the pollitos problem?

- Try using the lesson analysis tool to analyze the story and see what stands out to you.

Focusing on Knowledge and Identities

Centering Cultural and Community Funds of Knowledge of my students was not an abstract concept for me—I actively worked to know more about what they and their families knew and experienced because it enriched our learning. While I was also a Latina, I also knew that life for the children of Mexican immigrants in my city was different from my own and that each family was contributing unique practices and conocimientos that, together, made our classroom community one filled with knowledge. This was a problem that was very relevant to my students—they knew that pollito from the Pulga Ocho Doce, they knew how Kelis had been caring for it, they had seen

Kelis's mom bring it to our classroom in its carrier and join us at the rug before taking it back home, and they had a fond memory of all getting to meet and pet the pollito. The problem that day was something that everyone could immediately connect to and that demonstrated how I really did value the stories and experiences that they shared in morning meeting. The problem-solving space that we had created during that year certainly emphasized student thinking and ideas, and I had an explicit goal of *(Re)Humanizing Mathematics* in mind when it came to planning activities and experiences. Affirming my students' math identities was something that I explicitly centered in my actions in the classroom—from revoicing their ideas to emphasizing students' words, to ensuring that José saw his drawing valued and mathematized by his community. *Honoring Students' Thinking and Ideas* was a central focus throughout every stage of our problem-solving lesson. This meant that their words, their movements, their drawings, and their gestures all were given value, regardless of typical expectations for managing student participation or movement. Moreover, because my definition of translanguaging extends the welcoming of students' language practices beyond those that cross-named language boundaries to ensure that they are able to speak and share their ideas more freely, even when that means transgressing traditional boundaries for classroom talk, we can see (re)humanization in the ways students share out. They may be calling ideas out as they come into their heads during our discussions, jumping out of their seats to point at something on our charts, or approaching me as I work with other students to join in the collaboration. Affirming our students' identities requires us to think very carefully about the many practices in classrooms that work to stifle, silence, and suppress our students' abilities to express themselves.

Focusing on Rigor and Support

All students were invited to engage in thinking deeply about and making sense of mathematical processes and the underlying structure of mathematics itself. While I wanted students to "see" the tens in 117, I also knew that they had to think and make sense of how to see that in a way that held meaning to them. Students knew that, even if they could recite a correct answer, as AJ had, that they would also need to explain and justify their thinking and imagine ways of representing their understanding. All of this contributed to *Sustaining High Cognitive Demand*. To ensure that this expectation was also one that felt attainable regardless of students' comfort levels or experience with mathematics, I thought through supports across all stages of our lesson that could **Scaffold Up** for all learners: from the careful bilingual launch, to circulating and ensuring students were able to access materials or peers that would support their thinking, to actively engaging in questioning that could scaffold up for students unsure of where to start. In the discussion, language and visuals were used and connected to individual students' strategies so they might be able to see how their drawings or manipulatives connected to more abstract concepts. In particular, by re-creating a key part of José's work, with the tens and ones, I helped focus the class on the mathematical question at hand, which was "what did José do?" In this way, I provided support, in the moment, for their own reasoning to continue.

Finally, *Affirming Multilingualism* was always a fundamental aspect of how we engaged in mathematics. I knew that my students had only ever experienced math in English, and that, as a result, many of my students who felt more comfortable sharing their thinking in Spanish were rendered silent in math during kindergarten and first grade. I could see their silence as soon as the year started and all the math thinking I heard was in English. Even now, it is impossible to say what kind of a math learner Arturo might have been had his language been welcomed from the start. That being said, throughout this lesson, my multilingualism and the multilingualism of our class community were made normal—when Aliyah and Kelis shared the same idea in both languages, we all had access. When Arturo announced his answer in front of the whole class in Spanish, I made sure to repeat it in English so that everyone could hear his success. This was very intentional—it was part of the effort to (re)humanize mathematics by ensuring that it was also matemáticas.

Focus on Power and Participation

One challenge that I had when starting the school year was the expectation that, as the teacher, I would tell students whether they were doing it right or whether they had gotten the wrong answer. This meant that active efforts toward *Distributing Intellectual Authority* were a major focus of my efforts to build mathematical relationships with my students. In this discussion, José's drawing holds math knowledge, and it is students who will use their math knowledge to prove that. When Gabriel and Juan Carlos need help, it is Arturo's notebook that holds the key for making sense of the problem. And when Maricela and Chantel are praised for their calendar, Chantel reminds us that it only exists because "we helped each other." All of these examples also showcase how, when we teach mathematics in culturally responsive were ways that seek to recognize and celebrate the brilliance of everyone in the room, it becomes much easier to see how mathematics classrooms can be sites for *Disrupting Status and Power.* One reason I am so happy watching this video is because it captured such an important day in this year of math teaching. On this day, Arturo confidently voiced his mathematical idea during the discussion and used his strategy to support his peers, Maricela was named as someone who helps others with math, and Chantel reframed herself as a cooperative and supportive friend. At the same time, all of us worked on our year-long project to enact a mathematics in which language was not relevant to status or power in the classroom. It is important to note how much we need students to be a part of driving these efforts—we cannot *Disrupt Status and Power* by exerting our authority, we have to create space and seize opportunities to allow students to make our math classrooms into spaces of collective celebration and success.

While this problem-solving lesson is not one that connected to our units devoted to *Analyzing and Taking Action*, we were actively engaged in that work as a classroom. Just before we began problem-solving, we were adding to our jar for Flint. The jar had been the class's idea—they felt compelled to help people who they had learned were experiencing a hardship that felt more real because of how we had mathematized it. We had thought about what it really meant to use water bottles for nearly every aspect of our water consumption and the number of bottles that would require (including how different family members might require more water than others). We had even thought about how much water they would need in a school day—they decided on one and one-half sixteen-ounce bottles each. This led us to measure and discover halves and begin to understand fraction concepts. At the same time, we read articles, watched news videos, and learned about the effects of lead on the human body. For our classroom, making sense of how power worked to impact communities and what people could do about that was part of learning how to be active citizens and empowered community members.

PAUSE AND REFLECT

- What kinds of lessons have you seen inspire this way of thinking in other content areas?

AREAS FOR GROWTH

At the same time that I find much to feel proud of when reflecting on this lesson and the year of teaching that surrounded it, I wonder about the contrived nature of the mathematics in this situation. How might a modeling task have opened up the possibilities for students to think on their own of the elements involved in caring for a new pet, how might they have made different choices about feeding, and how might they have posed their own problems based on a more authentic engagement with the idea of caring for the pollito? While the experience was authentic, the problem that I posed was not—it very much took an experience and converted it into a more traditional school task. And the very nature of doing so meant that I still held the intellectual authority to determine what counted as a worthy task. This, of course, points to an inherent tension—how do we open up space for authentic student ownership of our classrooms and the learning that we do, while also ensuring that students are able to access, experience, and demonstrate their knowledge of a required and standardized curriculum? There are no easy answers to this, but, as a bilingual educator, my work constantly involved finding ways to navigate two competing realities and I always found comfort in embracing nepantla, or the space in-between, right where that tension lies (Gutiérrez, 2015). Ultimately, I know that, given the constraints of my specific context, I can look back at my classroom and see that my bilingual students were able to express themselves—mathematically and through language—and have their ideas, brilliance, and corazones recognized.

A FINAL NOTE

Melissa's lesson example helps us to see how the creation of a just bilingual classroom aligns with the creation of a just mathematics classroom. By using the CRMT2 tool to analyze a day in this second-grade classroom, we are able to see the various dimensions involved in creating bilingual mathematics classrooms that celebrate students' whole selves and look for ways of being active and participatory members of local and global communities. Despite the realities of structures imposed on classroom teachers, this chapter demonstrates that teachers have a tremendous amount of power to share with students as they work to build a way of doing math and doing school that honors all the corazones in their shared space.

This chapter shows that there is a special power in being a bilingual teacher of mathematics. Many bilingual teachers did not experience bilingual education as students. While there is always a tendency to replicate traditional math teaching practices in ways that can, at times, do a disservice to children, bilingual teachers have a unique superpower. The very fact of being bilingual teachers means they must create spaces that will inherently look and sound different from the way they experienced school as students. Because of that, they are called to reimagine what it means to teach, inviting their full selves and their students' full selves in ways that perhaps feel like something that only existed in sueños. They get to remake educaciones—el derecho y la obligación que viene con ser maestrx bilingüe. And remaking mathematics educaciones can and should be a part of that—to create the mathematics classrooms that make sueños into realidad and ensure that the ways that bilingual teachers redesign education allows for bilingual students to redesign matemáticas.

DISCUSSION QUESTIONS

- What purposes brought you to bilingual teaching? How can your math classroom work to serve that same purpose?

- What aspects of your mathematics classroom are working to (re)humanize your students? What aspects are countering those efforts? What does that mean you might have to ask yourself to do?

- If you do not identify as a bilingual teacher, how are you working to ensure that your classroom is an affirming multilingual space?

- Does your school/district make space for your bilingual students to be bilingual in their mathematics class? How might you structure an argument to make that change?

- How can you build space for students to make demands and take action within and beyond the classroom you share?

CHAPTER 9

CRMT IN SPECIAL EDUCATION SPACES

By Talya Kemper and Maria del Rosario Zavala

What does it mean to be a culturally responsive mathematics teacher to students with learning differences? Sometimes these students are included in our grade-level classrooms, and sometimes students are in self-contained classrooms. No matter the setting, culturally responsive special education teachers know that their students deserve access to rich, rigorous, and relevant mathematics, designed especially for them in light of their cultural and linguistic backgrounds. They learn deeply about their students' interests and capabilities, so that they can hold them to high expectations, with flexible and fluid ways of engaging them in learning.

Culturally responsive teaching and neurodiversity should go hand in hand but often do not. Within special education contexts, teachers usually adapt curriculum and instruction to individual student's unique learning circumstances, based on their Individual Education Plans (IEPs). However, deficit perspectives sometimes cloud the judgment of teachers of students with learning differences, leading them to underestimate their students' capabilities. In mathematics, deficit perspectives can dehumanize students—decentering who they are as people and focusing on narrowly defined metrics to measure their educational progress. Critics have been outspoken about how a medical model dominates mathematics instruction in special education spaces, where children are assessed in narrow ways, and prescriptions (aka, interventions) are spoon-fed to reach narrowly defined goals. This model is not compatible with Culturally Responsive Mathematics Teaching (CRMT). As Lambert (2018) argues,

> Students with LD (Learning Disabilities), as well as those who are labeled as low achieving in mathematics, are offered a more procedural mathematics when compared to their peers. The documented positive outcomes of an inquiry approach to mathematics, not

only in mathematical knowledge, but in developing student identi-
ties as agentic problem-solvers and a positive relationship towards
mathematics, mean that we can no longer allow such beliefs to
curtail the mathematical potential of students with LD. (p. 73)

CRMT offers a way to humanize mathematics instruction for children with learning differences. The co-authors of this chapter, Talya Kemper and Maria Zavala, wrote it based on our mutual interest in supporting teachers who work in both inclusion and self-contained contexts to learn more about how they can be culturally responsive with their mathematics instruction. We are colleagues within the California State University system and have had a friendship of many years. Talya is also a teacher in a local urban school district, and it is from her daily experiences across her long career as a teacher that we draw the vignettes in this chapter. Our combined perspectives from the mathematics education and special education sides illuminate some of the power and possibilities of CRMT with students who have learning differences.

In this chapter, we reconstruct and analyze particular mathematics lessons from Talya's teaching across two distinct special education teaching contexts, with a focus on two neurodiverse children. We then analyze the dimensions of the Culturally Responsive Mathematics Teaching Tool (CRMT2) at work within each lesson, focusing in particular on dimensions that show up most prevalently in each example.

The first context is a general education classroom in early elementary where a student with autism, whom we will call Mateo, received push-in support from a paraeducator. In this vignette, I (Talya) am the teacher, and the other adult working with Mateo is his usual paraeducator. The second situation is an upper elementary special day class, which is a small, self-contained classroom with students with severe disabilities. The focal student, Sunny, has cerebral palsy and intellectual delays along with other co-occurring disabilities. In each vignette, we provide some of the specific details around the focal child, a general description of the mathematics lesson with key teaching moves noted, and then a brief discussion of how we see CRMT at work within the lessons, in particular in relationship to the experiences of the focal student and, as in Sunny's case, the rest of the students attending the special day class. In this way, we illustrate how the CRMT2 framework can serve as a tool for reflection and analysis of teaching after a lesson.

LEARNING FROM MATEO

Welcome to Teacher Talya and Ms. B's classroom, a co-taught first-grade general education classroom. It is fall—about seven weeks into the school year. Meet Mateo. He is a seven-year-old boy in this classroom with nineteen other students. He is autistic, and in his case, autism impacts his communication, social skills, and academics. Mateo does not receive pull-out services, but instead he is fully included in the classroom and receives the support of a paraeducator, Ms. Temple. All three adults are often working collectively to meet the needs of different students.

Mateo has a severe expressive communication disability. He does not speak using his mouth. In order to communicate, Mateo uses an iPad with an app that helps him communicate. Mateo has had this communication system since preschool, and he uses it mainly to answer questions and request highly preferred activities. The app he uses is a communication system that family and staff must be trained on in order to assist students in using it. His family speaks primarily Spanish at home, and his communication device is programmed for use in both Spanish and English. As a highly proficient device user, Mateo has a large number of communication options on his device array. He loves reading, books, and anything related to video games.

In class, Mateo does not display much interest in interacting with his peers. He becomes upset by loud noises, people coughing, and people crying. Mateo has an IEP, and his goals include one specific to mathematics:

Mateo will receptively demonstrate (pointing, via choice) early quantitative concepts (e.g., "more" and "none"), qualitative concepts (e.g., "different"), and directional/positional concepts (e.g., "before," "in," "under") when presented in curricular contexts.

Recently, the children in this classroom have been working on addition and subtraction between 1 and 20. They are using mostly count-all strategies (where children can see all the objects they need to count and total them up using counting skills.) The problems are joining ("putting together") and separating ("taking from") problems, with the results unknown.

On this day, the class has a worksheet that has the top two examples written numerically and demonstrated with an image (e.g., an image of the number 4 with four apples, the plus symbol, an image of the number 2 with two apples, the equal symbol, and space to write in the number solution and draw the apples). The remaining examples on the worksheet are just the problems with no pictures, and students are expected to draw what they need to solve it. In all, students are making sense of symbols for addition and subtraction and the equal sign and using counting strategies to solve addition and subtraction problems (joining and separating.) Linking cubes are also provided for children to model the problems with manipulatives.

Given what the class is working on, Talya (the special-education teacher) has modified the material so that Mateo is working on a related topic suited to his independent communication level and known skills. While she did this in collaboration with Ms. B, Talya is also leading the lesson on this day. In this case, Mateo is working with numbers between 1 and 20 and understanding the concept of relative quantities (more, less, equal), and so his device is set up with related communication buttons available (see Figure 9.1).

In this moment, Mateo is working one-on-one with his paraeducator, Ms. Temple. As he shows what he knows through his communication app, today he is getting a lot of use from the *more* and *less* symbols as Ms. Temple asks him to compare the quantities in each problem. For example, she asks "Is 4 more than or less than 2?" And Mateo then pushes the button for "More."

FIGURE 9.1 Mateo's Communication Device

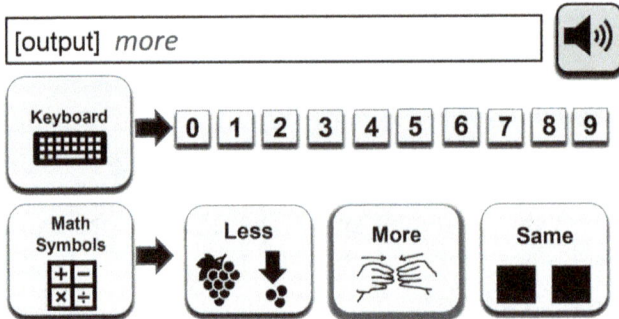

This is a simplified re-creation of Mateo's communication app, where he navigates between the keyboard and the math symbols buttons to share his ideas.

In this way, in this part of the lesson, Mateo is working on a related aspect of addition and subtraction for numbers 1 through 20, while targeting his IEP goals. He is also using counting strategies to compare numbers and making sense of symbols, but he is not yet working on the addition and subtraction portion as he works with quantities, which is where his teacher would like him to go over the next few days.

The class is getting loud as they get into their work. Mateo has headphones on to help him focus. Talya circulates and comes over to Mateo. When Talya comes over to sit with Mateo, the paraeducator takes over the role of circulating and supporting other students. But first, she fills the teacher in on how Mateo is doing. She shares that Mateo experienced some struggle with this activity, and his responses had only a 50 percent accuracy.

Talya sits next to him and places the worksheet the class is working on in front of him. She is thinking about what the paraeducator told her and how it may mean any number of things: *maybe Mateo is bored, maybe tired, or maybe Mateo is still working on the concept.* Talya needs more information. She is there to elicit Mateo's mathematical thinking. She plans to do this in at least two ways; once through asking him to point to numerals and once asking him to use the blocks to support his mathematical thinking. She hopes these two ways help Mateo learn that comparing numbers is attached to specific quantities, a concept he may be familiar with but is worth repeating because soon he will use the blocks to do something new: model addition and subtraction. So she asks him to point, instead of using his communicative device, just for these questions. This is also social time, time for her to deepen her relationship with Mateo and build his confidence. Talya keeps her tone positive and supportive.

"You aren't too tired yet, are you?" she asks. Mateo looks toward the little classroom library area where he usually takes his breaks. Talya mentally notes he might be getting tired but decides to see what he is up for first.

"Remember how much counting you did yesterday? You were on fire!" She gently teases Mateo as she places some cubes from the table between herself and Mateo. She makes two piles: one with five cubes and one with three cubes.

"Okay, Mateo. Which one is more? Is the five more, or is the three more? Which is more?" As she says each number, she points to it. She does it one more time, and Mateo follows immediately by pointing to the five. "Yes, great work. Let's try one more, in this problem, eight plus three equals something. I don't know what it equals yet. But let's start with which is more, eight or three?" She says the last part slowly, leaving her finger pointed to eight for a moment, then leaving it pointed to three for a moment. Mateo points to the eight. "Okay, great work on that!"

She tries one more problem. She tells Mateo there are four on the left, counting them aloud as she tags each one. She puts six counters on the right and counts them aloud. She then asks Mateo, "Which is more?," indicating by pointing, "the right or the left?" She prompts Mateo to point for her, and he points to the right. "Yes," she says, "the six blocks are more than the four blocks. Great work, Mateo!" At this point, she is encouraged by his repeated success, but she doesn't stop yet. She is going to do a few more problems but remove the scaffold of counting aloud for him and instead leave the whole solution up to him. She asks him to count and puts a laminated piece of paper with the numbers one through ten printed on it, prompting Mateo to use the chart to tell her how many are in the first pile and then the second pile. Then she asks him to indicate which is more by pointing, just as before. Mateo is successful three times in a row, with three different comparisons. He counts all sets accurately, and he indicates by pointing which is more. "Good job, Mateo! Tomorrow we will do addition, and it's going to be fun!" The teacher makes a mental note to have Mateo work with a small group of students for center work the next day, already thinking about which kids would be a good small group for Mateo and the paraeducator to work with.

Mateo puts his device in his desk and indicates with a particular hand movement that he is done working right now. The teacher reminds him he has a couple more to do. Mateo shakes his head no and also waggles his finger at the teacher, his communication for "No, I need a break now." Talya understands, and says, "Okay Mateo, how about a two-minute break?" She gestures to the paraeducator, who then goes with Mateo to the library area, where he lays down and looks at books while she sets a two-minute timer on her phone.

A student waves Talya over. She says that it's not fair that Mateo gets to lay down. Talya takes the opportunity to remind the student that we all need different supports to be successful, and it's important that we give each other what we need. Next, she asks the student what she is working on right now and how it's coming along. She then asks her what she needs, and the student asks her a question about a problem. They chat for a minute, then Talya encourages her to work with her neighbor and leaves them to keep figuring out the problem with their blocks.

Noticing the time, she tells the class it's time to put the work they are doing in their math folders on their desks, clean up their tables, and line up for recess. She reminds them they will have sharing time after recess to go over the problems and learn what strategies people used. The children prepare for recess.

ANALYSIS OF MATEO'S VIGNETTE

Let's take a peek into the decision-making that Teacher Talya is engaged in, especially as it concerns Mateo. The first instructional decision is the assignment for Mateo. There are modifications for how Mateo will do the work and how he will use the same task to work on something mathematically different but related. One thing we can say is that Talya is simultaneously working on his defined IEP goals and his grade-level content needs. While she works with Mateo, she needs to gather evidence of both: his goal progress and his first-grade mathematics learning progress. Certainly, at times this evidence overlaps. For example, his work with comparing quantities with more or less gave him many opportunities to use his counting skills. This gave Talya an opportunity to see how they were developing so that they can be used to solve joining and separating problems next.

In relation to the CRMT2, we'll focus in on five dimensions we think most prominently feature in this vignette: *(Re)Humanizing* (2), *Student Thinking and Ideas* (3), *Cognitive Demand* (4), *Scaffolding Up* (5), and *Affirming Multilingualism* (6) (see Figure 9.2). Because in our analysis of the vignette the other dimensions did not show up as prominently, we did not include them in our discussion.

FIGURE 9.2 CRMT2 for Mateo's Vignette

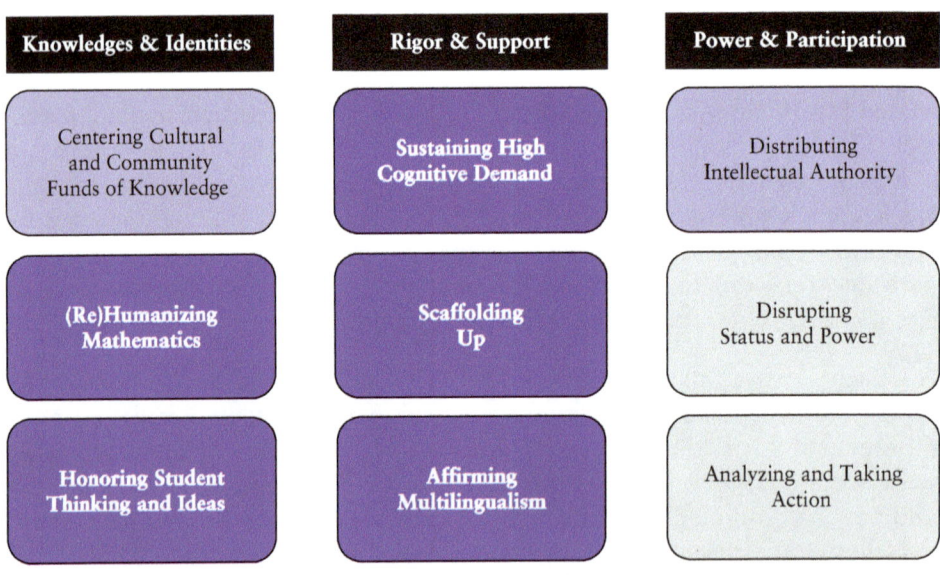

The dimensions analyzed in Mateo's vignette are in dark purple. The dimensions Talya wants to work on next are in light purple.

(Re)Humanizing Mathematics (Dimension 2)

First, we see evidence of *(Re)Humanizing Mathematics* in the classroom as Mateo and the teacher work together. From the first moment she interacts directly with him, she moves beyond thinking of him as a number (e.g., "50 percent response rate") and focuses on probing his thinking. It was important that she moved beyond the single way of doing math, being "right or wrong," and instead worked with him using manipulatives, to be able to observe his approach to the problem, while she monitored his solution methods. While the teacher is well aware of a need to document evidence that Mateo is moving toward his goals, she is also well aware of his abilities. She expects him to be able to participate and perform at a high level, for him, and that, even if he needs a break, he can continue to be an active participant. Essentially, we see affirmation in Mateo as a mathematical person, able to reason and express his reasoning. While this particular lesson did not have opportunities for Mateo to creatively express his ideas, Talya can note that and ensure that as the class continues to work on open-ended word problems, she will be able to engage Mateo in some modeling of addition and subtraction and explore the ways he expresses his thinking with manipulatives, gesture, expression, and his communication devices. In this way, she also attends to how being a mathematically competent person does not mean something narrow, such as accurately showing paper and pencil reasoning but rather that mathematical competence can be expressed in multiple ways. We note together that some aspects of humanizing come through stronger than others in this lesson, but overall there is clear evidence of rehumanizing at work in the interactions in this lesson with Mateo.

Honoring Student Thinking and Ideas (Dimension 3)

As part of how autism impacts Mateo, he is learning that when he shares his thoughts, other people can know what he is thinking, and when he doesn't share his thinking, people don't know what he is thinking. Therefore, he is given a lot of support to communicate his internal thoughts through his devices. Compare his development to what might be going on for a child who is not on the autism spectrum who is learning to take their inner thinking and narrate it—Mateo will not be narrating his thoughts in this sense but instead is learning that he can share them in other ways. This is a different skill set that is developed explicitly and then through the math lesson when Talya models talking aloud about her thinking (i.e., counting the objects in the set, explaining how she knows which is more) and is part of how Mateo is invited to also learn to express himself.

Talya's pivot from focusing on Mateo's response rate to eliciting his thinking is an important move to notice. After she observed him solve the first two problems, taking in how he used gestures and pointing to numerals to communicate his ideas, she had a good sense of what he was doing to solve the problems. She also wanted to observe (or "listen") more closely, so she provided multiple opportunities for him to demonstrate his thinking. Because she elicits his thinking, she can also respond to it. She both praises his effort and helps him recognize his accuracy, but she also gives feedback that because of his developing skills he will be able to engage in more challenging mathematics ("Tomorrow we will do addition and it's gonna be fun!").

Sustaining Cognitive Demand (Dimension 4)

As described in Chapter 4, *Sustaining Cognitive Demand* is about the extent to which children get to engage in exploration and analysis of mathematics concepts, procedures, and problem-solving strategies. In this case, Talya, in her role as teacher, chose to hold Mateo to the same level of cognitive demand that she had the other children. He was expected to be able to reason through his math problems. We note that this dimension could be stronger if she had asked Mateo to answer the question "How do you know?" more often, such as "How do you know six is more than three?" However, she also has to consider keeping Mateo engaged across the duration of math time. At that time, she knew they would be doing more mathematics after recess, and she wanted to ensure she was not burning him out in the first part of class. Since the children would be sharing their strategies for solving the addition and subtraction problems after recess, she wanted Mateo to focus his energy on hearing and understanding the strategies of his classmates.

Scaffolding Up (Dimension 5)

One of the trickiest spaces that special education teachers work in is meeting the IEP goals of children while ensuring they are engaging the children in rigorous mathematics. Even as Talya supports Mateo to reach his IEP goals, which are written as measurable and attainable, she also must engage him in deep and rich mathematics. We see strategic scaffolding at play in Talya's interactions, as she works with Mateo on the concepts of comparing quantities using mathematical language of "which is more" and "which is less," building his quantitative reasoning skills that will lead to solving problems involving addition and subtraction. She is using the activity he is capable of to build his numeracy in ways that will be beneficial beyond comparing numbers but also for combining numbers and then even taking a smaller number away from a larger number. In doing so, she is helping him meet his IEP goals while thinking carefully about scaffolding Mateo into the harder work. It is important to note that she is planning to get him there, not abandon him with one skill gained in isolation.

One specific planned scaffold was to give Mateo a modified worksheet where he was able to use the top two examples with images to practice which was "more" and "less." After the first two examples with the images, Mateo and Ms. Temple moved on to a worksheet made by Talya (his special education teacher) in collaboration with his general education teacher. Having gathered more evidence today that Mateo uses one-to-one correspondence to count accurately and consistently, and that he can tell the difference between more and less, Talya plans to move him into totaling up two sets of objects and introduce addition as joining as his next objectives. In this way, she is also moving him closer to doing the same mathematical activities his peers are doing.

Affirming Multilingualism (Dimension 6)

As a bilingual child with learning differences, Mateo has many factors impacting how he interprets a mathematics problem and produces an answer, including understanding the language of the question, what is being asked mathematically, and what form the answer should be in (which in his case could be numerical, the selection of a group, or something else he communicates with his device). For this reason, Talya needs to employ many strategies to affirm multilingualism. By incorporating gestures, sign language, and other forms of communication, such as through his device or through manipulatives, multiple modalities of communication are affirmed. Talya does a lot of revoicing of Mateo's ideas back to him, both as a form of affirmation and to model language usage. This is also the way she affirms his contributions. She uses manipulatives not just as a tool for problem-solving but also as a tool for communication and thinking. By using the cubes, Mateo has a tool to make his thinking visible and demonstrate how he is counting to compare quantities, and he and the teacher have a tool to interact around. Through these strategies, Mateo's identities as a multilingual learner are centered throughout the lesson.

Reflecting on What's Next for Mateo

As Talya reflects on how she supports Mateo, she can note that some dimensions will need to be developed in other lessons more—such as drawing explicitly on funds of knowledge and attending to power and participation. His *Funds of Knowledge* (Dimension 1) can come into play more as she develops joining and separating problems for him to solve, leveraging what she knows about his interests and experiences. She can also attend more explicitly to *Distributing Intellectual Authority* (Dimension 7) as she considers how to position him as a learner among his peers during a strategy-sharing discussion. For example, she thinks about how she might plan to assist him in sharing his solutions with his classmates once he works them out. While we may see how some of these dimensions began to emerge in the vignette, Talya's concerns are with the whole school year and how she maintains her classroom as a space where every student can feel connected to mathematics and valued for what they know. Therefore, she can work on strengthening those dimensions in future lessons.

For her own next steps, Talya reflects on the following: Mateo had in previous lessons mastered successfully ordering numbers 1 through 20. His difficulty with the concepts of more and less may point to a number of factors. For example, his learning to order numbers 1 through 20 was rote memorization without an understanding of the concept; or it may have only been an issue of his receptive English language skills; or it may have been a mechanical issue of continuously learning how to navigate his communication device; or a combination of things. Particularly when teaching and assessing Mateo's understanding of math concepts, it is necessary to consider how his understanding is impacted by his emergent bilingual language development and by learning to use his communication device and software. This is why Talya looks for many opportunities for Mateo to demonstrate his understanding and plans to provide him a more challenging task, contextualized in his love of *Pete the Cat* books (e.g., *Three Bite Rule*). She will continue to assess where his strengths lie and troubleshoot any issues with his communicative device, as well as gather evidence if there are emerging language issues that she needs to accommodate for. The problems she plans to give him will draw on his experiences and interests, in this way bringing **Community and Cultural Knowledge** (Dimension 1) into the lesson. And she will check in with him as they solve problems to see if he's up for sharing his strategy, with her support, to the class (**Distributing Intellectual Authority**, Dimension 7). As she makes her plans for the future, Mateo's needs as a mathematics learner are central to her decision-making.

LEARNING FROM SUNNY

Welcome to Teacher Talya's K–5th special day class for students with complex communication needs. Special day classes are where Teacher Talya has spent most of her career. In this classroom, she has six students, two paraeducators (paras), a custodial aid, and a nurse who does, among other duties, the feeding of students since they have G-tubes (the short name for gastronomy tubes that deliver nutrition directly to the stomach for people who have muscular issues that prohibit them from eating). Sunny is an eleven-year-old Vietnamese boy in fifth grade. He has cerebral palsy, an intellectual delay, is hard of hearing due to severe hearing loss in his right ear, has a G-tube, and is medically fragile due to a host of co-occurring issues. His disabilities impact his communication, social skills, and academics.

Like many students in this dense urban area, Sunny lives in a small apartment with his family. His family speaks Khmer at home and Sunny understands both English and Khmer. Sunny loves any activity that involves music and will move his body and smile with absolute joy when an activity involves music. His limited mobility and limited family resources mean that his experiences are mostly in his immediate neighborhood. His neighborhood happens to be a bustling and busy part of the city, close to parks, shops, murals, and other features of this dense urban area.

Sunny has a host of medical issues that Talya has to consider when designing instruction. He is sometimes sleepy in class because of his antiseizure medication. When he is not sleepy, he is an active participant—making noise, spontaneously laughing, and enjoying himself. Finally, he is not able to see very well, and so brightly lit objects hold his attention far better than anything else in the class.

Sunny is emotionally expressive—quick to laugh when something is funny and will cry if overly frustrated. However, he can only reliably move his head and does not have much other muscle control. Because of this, manipulatives have to be large and easily gripped so that a teacher can place it directly in his hand and help fold his fingers around it. He is not able to speak using his mouth. In order to communicate, he blinks his eyes to indicate yes and shakes his head side to side for no. When Sunny looks at her, smiles, and laughs, Talya knows he is engaged and ready to participate. When Sunny is disengaged, he can look like he is sleeping or sometimes actually falls asleep. He cries or shifts in his seat if he has explicit bodily needs that need to be addressed (such as when he wants to lay down or needs to be changed).

All of the students in the class have a severe communication disability and extensive health needs, some similar to Sunny's and some unique. None of the students are ambulatory, and they all are working on communication skills integrated with academic learning. In addition to academics, there are many other competing needs occurring in the classroom that must be juggled around academic learning. However, mathematics is a key part of the instructional day. Talya is committed to ensuring the students get a multidisciplinary curriculum, including reading, writing, physical activity, mathematics, and life skills.

Sunny has an IEP; he does not have a goal related to math or academics, but he does have what are termed "pre-academic" and expressive language goals. The goals that link to a math activity are answering a yes/no question, participating in an activity, and being able to label differences in objects (e.g., one is big, one is small).

A key way that teachers work with Sunny is through field of vision and choice of object, then verbal communication to confirm the choice. When Sunny is shown different objects or given a story problem, he must indicate from a field of two images what his answer is. For example, if given the question "Six children stayed home, and two came to school. Where are there more, at home or at school?" Sunny will independently use eye gaze to indicate a choice from a field of two icons or pictures and subsequently confirm his choice using a head nod. Sometimes the images are the amounts. Sometimes they are the categories. Communicating one or the other choice is a primary way Talya engages the students in math activities.

Talya demonstrates how she might put these images up and hold them in her student's field of vision to help students answer "Six kids stayed home today, and two are at school. Where are there more kids, at home or at school?" (see Figure 9.3).

FIGURE 9.3 Image of Board With Two Pictures

Today for mathematics, the students all sit in a circle with trays on their wheelchairs in order to place materials on them. Communication systems are also available, as needed. The two instructional aids, Ms. Pham and Ms. Tyler, sit behind the students during instruction, moving to each student as needed. Talya sits in front of the students in a rolling chair so that she can reach all of the students. She uses a Promethean board to put the lesson on. The lesson today is focusing on the concepts of "more" and "enough." The math lesson uses manipulatives and visual aids, and the problem is presented in a story format. It is displayed in large font on the board and she reads it aloud; the story features a well-known teacher, Ms. B:

GLOWSTICKS TASK

Ms. B brought eight glowsticks to school. She wants to give one glow stick to each student. There are six students in school today. Does Ms. B have enough glowsticks for all of the students?

After reading the story Talya then brings out the surprise—actual glowsticks!

These are thick glow sticks, which allow for the children to hold them, some with adult support and some with their hands resting on their trays. She fans the glowsticks out in her hands and lifts them up, showing them to everyone in the class. Their glow is far easier to see than other manipulatives she might have used. Sunny can see them. She hears his excitement in the noises he makes. She says to the class, "What should we do with these glow sticks? Do you think we will have enough for everyone?"

She then goes around the half-circle of students, eight glow sticks fanned out in her hands, and gets an answer from each student, marking on the board how many votes there are for each answer: yes, we have enough, or no, we don't have enough.

Talya reaches Sunny and asks if there are enough glowsticks for everyone. She tells him he can blink for yes or shake his head back and forth to say no. She tells him to listen to his two choices and reminds him how he can answer: "You can blink for yes, we have enough for everyone, or shake your head no, we don't have enough." Talya also reviews the visual aid a few times with him, fanning out the glow sticks and circling them through the air so he can enjoy the movement of dancing lights. Holding them steady for him to think, Talya asks Sunny if he thinks he should blink for yes or shake his head for no. Sunny blinks his eyes indicating there are enough glow sticks.

When all students have shared, she calls their attention to the tallies she drew. "Okay, which one has more?" It is one more opportunity to work on more or less concepts. There are four tallies for "yes" and two tallies for "no." Talya circulates again and also listens in as the two instructional aids repeat the question to two other students. For the moment, Sunny sits by himself, immersed in the general murmurs of the class, his gaze toward the glow sticks.

Talya gives the instructional aids a moment to get an answer from their students. She observes that we seem to have consensus that there are more yes responses. "Ah, it seems like this one won. It seems like we think we have enough. Four people say yes, that's more than two. But now, let's see!"

At this point she moves around the room handing a glow stick to each child, under the close watch of the paras since one student is known to put everything in her mouth. She gently opens Sunny's hand and places the glow stick in his palm and talks through the situation as she circulates. Once the glow stick is in his hand, Sunny laughs.

"Do we each have one? Are there any left over? Let me see, I have two left over. So does that mean we had enough?"

After a brief pause she continues: "Yes, it does! I can even give glow sticks to Ms. Pham and Ms. Tyler," she says, dramatically giving each instructional aid a glow stick. They laugh and show them to the students. The adults show the children how they can wave the glow sticks around and have a little dance party. Talya goes to Sunny and takes his hand with the glow stick, moving it up and down gently. "Get into the glow stick party!"

After a brief glow stick party, students and paras put the glow sticks on each child's wheelchair and swap colors with children so that they have their favorites. The paras keep their glow sticks, so that each person still has one, demonstrating fairness that the children value (i.e., no one child has more than one glow stick). As the day goes on, they go about the rest of the day with their glow sticks close by.

ANALYSIS OF SUNNY'S VIGNETTE

As with Mateo's vignette, here we will discuss and analyze the lesson with Sunny through the lenses of select dimensions of the CRMT2: *Centering Cultural and Community Funds of Knowledge* (1), *Honoring Student Thinking and Ideas* (3), *Scaffolding Up* (5), *Distributing Intellectual Authority* (7), *Disrupting Status and Power* (8), and *Analyzing and Taking Action* (9) (see Figure 9.4).

FIGURE 9.4 CRMT2 for Sunny's Vignette

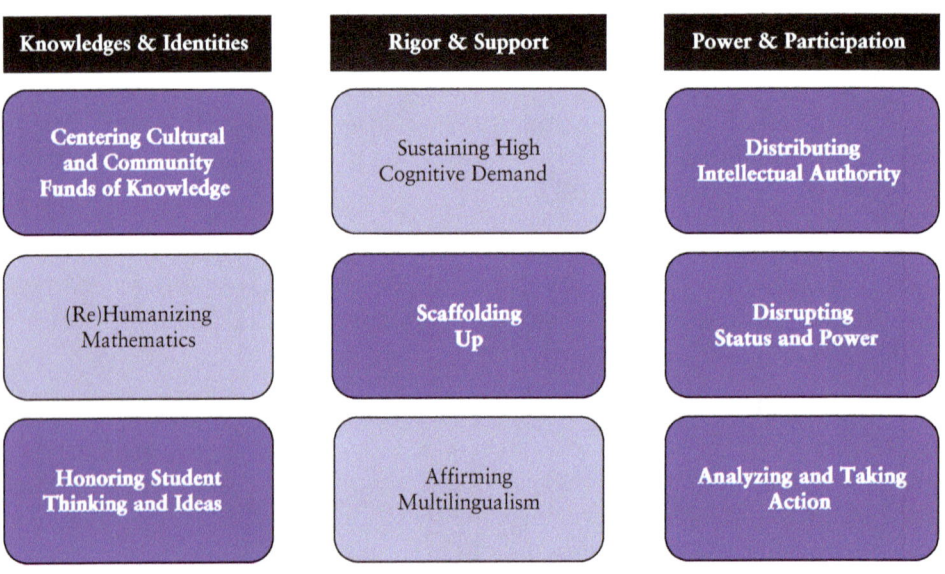

The dimensions analyzed in Sunny's vignette are in dark purple. The dimensions Talya wants to work on next are in light purple.

Centering Cultural and Community Funds of Knowledge (Dimension 1)

Talya considered multiple aspects of Sunny's and other students' funds of knowledge as she planned for mathematics instruction today. Sunny's cortico-visual impairment means that a bright, glowing object is more visible for him and students with similar visual impairments in the classroom. And Talya knew that Sunny, in particular, would love glow sticks and a dance party with V-Pop music, so she ensured that the math problem centered on his experiences for today. She also centers his needs in that she selected a manipulative that he could hold, as opposed to traditional mathematics manipulatives that would be much harder for him, and other students in this class, to actively participate with. Ms. B, a beloved teacher of the students, is also featured in the problem, adding to the connection to the context. All of these aspects were considered by Talya as she prepared her activities for the day.

Honoring Student Thinking and Ideas (Dimension 3)

Talya takes the time to ensure that every student's thinking is captured in the classroom. She indicates a tally mark on the board for each person's idea, so everyone gets to express if they think yes, there are enough, or no, there are not. The move to represent each child's thinking, and then to test out if they are right, by physically modeling the situation, as opposed to using her own authority to decide if they are, is also some evidence of Dimension 7: *Distributing Intellectual Authority.* In Talya's lesson, the students held the authority to both decide what is right and then work out if the solution is fair.

Scaffolding Up (Dimension 4)

As Talya plans for instruction, she thinks about what Sunny and the rest of the students need to be able to engage with the mathematics. One specific scaffold is the selection of the manipulatives. The glow sticks in the shape and size that the children can handle are one specific scaffold into the mathematics, because they provide a way to visualize the situation and allow for the class to model the problem together. Talya also uses the Promethean board to have the problem displayed as a reference. She physically records the response from each student so that they can count the tally marks to compare them. And finally, she prepares her questions for the students to allow them to communicate their answers as yes or no. It is evident that she planned multiple supports to address the needs of Sunny and his peers and that they permeated the whole lesson, not just one phase.

Disrupting Status and Power (Dimension 8)

Talya carefully attends to multiple ways that students are active participants in the classroom. One aspect of the access to the lesson is through attention to manipulatives used. Another is the expectation that every child will give an answer. This is significant in that it disrupts the entrenched stereotypes that children with physical differences and learning differences such as those in this classroom would not, or could not, participate in a mathematics lesson asking them to evaluate quantity and determine what the meaning of "enough" is in this case. Upon reflection, Talya shared that there was a moment where she asked herself if the time it takes to ensure every child has an answer was worth it, but ultimately she decided it was. It was a choice to not give in to pressures to rush through the activity. By reminding herself in the moment that yes, getting every child's input is time worth taking, she is also holding space for students with disabilities to be authorities in the classroom.

Analyzing and Taking Action (Dimension 9)

While this is not necessarily a social justice lesson, by focusing on the question of if there are enough glow sticks, Talya is talking about fairness. The question of "enough" is understood by the students to mean that each person gets one, which is fair. In this way, mathematics is also being used as a tool to analyze a real situation. Whether or not the students make this connection explicitly is unclear from the vignette, but Talya's impression as the teacher is that her students have a strong sense of what is fair to them, and fair treatment is prized among many of her students, especially when it comes to sharing special items that they like, such as glow sticks. Their analysis of whether there are enough is then carried out, and the students, along with the teachers, take action to determine if what they did was fair.

In Talya's view, fairness in this class seemed to be about being included and not being left out. Talya wonders if students would react differently if more than one glowstick was available for each person, like a classic division problem. But with the students, what seemed to matter most was having at least one. In particular, Sunny is a student who would cry or express distress if he didn't have one, but it's not clear if he would care if he didn't have more or if others didn't have one. Fairness was very personal for Sunny. We wonder if Sunny noticed that it was not fair if everyone didn't have one but also wondered how he would express that. That is the point to which Talya took this lesson: Fairness is when everyone had a single glow stick. But she reflects on how important repeating lessons like this both tap into students' sense of fairness and allow space for ideas of fairness to change over time.

Reflecting on What's Next for Sunny

For Talya's part, the next steps are to ensure that she continues to do lessons where every child has a functional way to communicate and staying abreast of changes to students' medications and body states, to make sure they can partake in the lessons. This constant checking in about body state is a normal part of instruction and adds a layer of responsiveness to students in a classroom like this. This goes back to presuming competence and not presuming incompetence, instead seeing your job as setting high expectations, giving tools to meet them, and then ensuring you as a teacher are attending to when they are communicating needs. She wants to do more with figuring out when a quantity is enough and explore Sunny and the rest of the class's sense of fairness. The reality is that there are academic goals. The students are not there to just sit all day, but they are there to learn, and that includes rigorous, rich, and relevant mathematics.

REFLECTIONS: "WAKES AND WAVES" IN THE MATHEMATICS CLASSROOM

After discussing the teaching vignettes together, we (Talya and Maria) found ourselves reflecting on both the opportunities and constraints of special education spaces. On the one hand, because of the nature of an IEP, many students receiving special education services are receiving tailored curriculum with multiple opportunities for it to be responsive to their communities and knowledges. However, as Lambert (2018) and others have pointed out, most of the time children do not receive rich problem-solving opportunities or even cognitively demanding instruction, *at their level*. Rather, they are instructed with a medical model, where a teacher-proof "treatment" is delivered that targets the learning goal. This is akin to saying that if the diagnosis is lacking understanding of basic addition, then the treatment is a packaged direct instruction intervention, regardless of who children are or what they are like. The medical model for mathematics instruction is ultimately dehumanizing. Therefore, *rehumanizing* takes center stage when teaching mathematics in special education contexts.

At the same time, we found ourselves asking, must IEP goals dictate the mathematics engagement of students? They are a reality of the context of special education spaces, but they are also challenging to write in ways that are meaningful to student learning. Since they must be measurable and quantifiable, as defensible to a legal standard, they oversimplify learning. IEP goals are also usually written with the idea that students develop incrementally, moving along clear, often linear, developmental trajectories, when in reality students may move in leaps, may appear to not retain knowledge, and then later seem to have figured new ideas out. In essence, learning isn't linear or incremental, not for all the neurodiversity of children. Even so-called typically developing children do not learn linearly. Why should we expect children in special education spaces to learn that way?

Talya is also a highly experienced teacher in what is termed the mod/severe (i.e., low incidence) context, having taught for most of her years in education in a self-contained classroom. In a classroom with low-incidence students, mathematics and science typically take a back seat to reading and life-skill goals. But as has been observed by a number of teachers that we have met in our work with special education credentialing, life skills and literacy have a lot to do with mathematics. For example, Maria's special education credential students working with low-incidence populations both in school and in continuing adult education have noticed time and again that mathematics skills related to budgeting and shopping, cooking, navigating the city streets, and other daily activities could be central to mathematics instruction in these special education spaces. But they are often not, because of the presumption that mathematics is too hard for these students, rather than finding out how students would or could use mathematics when given the opportunities to learn it.

If we genuinely believe in the humanity of all students, in light of neurodiversity, cultural and linguistic backgrounds, modes of communication, medical conditions, and IEP goals, then we teach mathematics in a culturally responsive way. This is what it means to view mathematics as a human endeavor. It's not extra-human, it's plain human. Mathematics is not less important than reading; learning mathematics is part of a well-rounded life.

And since we are teaching mathematics, then we must teach in ways that humanize all children, in all spaces, in every way possible.

Tan et al. (2022) use the metaphor of wakes and waves to describe the landscape of mathematics education for students with learning differences over the last few decades. They characterize the majority of the field of research into mathematics and special education as being full of waves that knock us off course—what they call "dehumanizing waves"—but note that there have been some bright spots of "humanizing wakes." Humanizing wakes are characterized by focusing on rich and rigorous mathematics, seeing students with learning differences as producers of mathematical knowledge, and critiquing systems and structures that contribute to deficit perspectives of children in special education spaces. CRMT can be a way to help educators ride high in a humanizing wake that genuinely values and centers all students, pulling students with learning differences from the margins to the center of rich, rigorous, and relevant mathematics instruction.

DISCUSSION QUESTIONS

- How do you ensure that your mathematics instruction for students with learning differences humanizes them and provides access to rich and rigorous mathematics *in light of* (not in spite of) who they are?

- Reflect for a moment on the interactions Talya has with her students in the vignettes. What strikes you about her teaching?

- What aspects of CRMT will you seek to develop and address in your own classroom spaces with students with learning differences?

EPILOGUE

empathy is the thread woven through patterns of speech
remembering imagined threats of ridicule and disdain
resolved to open windows
sealed behind false bravado
fear wearing the rugged individual
like a tent with bootstraps so strong they pull themselves up

victims no longer silent
voices found
becoming advocates for what is fair and unjust
in a system
weighted in favor of
the privileged, pre-schooled two-parent tradition

the teacher's heart beats
the memories of all the feelings
of acceptance and alienation
of inclusion winning over exclusion
of rushing from the margins
to the center of our attention.
we are all welcome here!
—Excerpt from "The Teacher Space" by Norm Mattox[1]

---

[1] "The Teacher Space" appears in the book *Black Calculus* (first published by Nomadic Press, 2021), by Norm Mattox. This excerpt is used with permission. For over thirty years, "MathMattox," as he was known to students, was a middle school mathematics teacher. His poetry is available at https://www.mathmattox.com/about

Welcome to the end of the book and the opportunity to take your next steps into your journey cultivating mathematical hearts through Culturally Responsive Mathematics Teaching (CRMT).

Throughout the book, we have asked you to pause and reflect, partake in a variety of activities, and overall go deeper into your own beliefs and practices. But this book is a starting point, not an end. We hope you have flagged pages to return to, scribbled in the margins, and discussed ideas with friends, colleagues, and family members. Perhaps you have worked with a group and used the Book Study Guide to work your way through the content.

We hope you feel energized and ready to start weaving a new fabric for your own mathematical practices that will foster joy and justice in how children experience learning mathematics in elementary classrooms.

In Part 1, we took you through a vast and dense braiding of ideas. We introduced the Culturally Responsive Mathematics Teaching Tool (CRMT2), its strands, and the dimensions that go with each strand. We introduced essential questions that help us reflect on our practice in relation to each dimension, ideas for approaches to instruction, and described the rubric for the dimension to serve as a guidepost for reflection and growth. To illustrate key ideas, we shared a teaching story. Along the way, we suggested a variety of resources to go deeper and cultivate your own mathematical heart. We hope you take full advantage of the activities and resources to continue your own learning about topics that you recognize you need more work in, as well as topics that are strengths you can build upon. But we are sure we also missed some good additional ideas out there. Let's keep the conversations going and, in the spirit of this book, look to share new resources and ideas with each other, strengthening our mathematical heartwork.

In Part 2, experienced teachers and coaches gave us a glimpse into how CRMT teaching can be an everyday practice, with key jumping-off points for centering children and their experiences. The chapters span general education, bilingual education, and special education spaces. The teachers who contributed to our book are indeed phenomenal. And, they are part of a vast collection of teachers who have unequivocal faith in the children in their classrooms, who strive to work in community with families, and who organize at the grassroots level for change in mathematics education. Together they deliver on a promise of mathematics instruction that is centered on the humanity of children, while critically reflecting on their own growth and development. All of us who wrote this book, who read this book, and who contributed to the book (children, too!) are part of this movement now. And we pledge to sustain it.

At times, our own growth will be painful, uncomfortable, and trying. And at times we will get fatigued. We will tire of struggling within and against a system that has so carefully dehumanized the children and families of our hearts. When we are tired, when we are dejected, we will remember what is at stake. We will remember and remind each other that each child deserves to shine. Every child deserves full time in the sun, out of the darkness. In her book *Unearthing Joy*, Gholdy Muhammad (2023) wrote, "Can you imagine if the sun shined only on some of our children and not others? To say that only some students are genius implies the sun shines only for them" (p. 19). In mathematics education, what is at stake is every child's time in the sun, every child being recognized for the genius they already are, and the mathematical hearts they already have.

> *"In mathematics education, what is at stake is every child's time in the sun, every child being recognized for the genius they already are, and the mathematical hearts they already have."*

And when we are tired, we will need to reach deep into our own resources and reach back into our collective energy. We will help sustain each other. Because we know that, ultimately, we have the responsibility to be the change we want to see in the classroom and in the other educational spaces in which we work. When it comes to cultivating the mathematical hearts of the children with whom we have the honor to work, they are the "flowers" in our midst. Muhammad (2023) reminds us that ensuring every flower grows is our responsibility: "The growth of a 'flower' calls for work from the educator—the one who is nurturing the flower to grow. It calls for us to cultivate—which is a social action and responsibility. We are the ones who care for the flowers. We are the ones who water their genius" (p. 19).

We hope this book helps you on your continued journey to cultivate your own mathematical heart, which is intricately connected to the mathematical hearts of others. Remember it is only through our connections, our braiding and interweaving of our hearts with others, that we make and sustain the change we want to be.

....

CULTURALLY RESPONSIVE MATHEMATICS TEACHING TOOL (CRMT2)

STRAND: KNOWLEDGES AND IDENTITIES

1. Centering Cultural and Community Funds of Knowledge (CFoK): *How does my lesson help students connect mathematics with relevant/authentic issues or situations in their lives?*

Fragile/Margin		Strong/Centered		
1	**2**	**3**	**4**	**5**
There is no evidence of connecting to students' cultural funds of knowledge (parental/community knowledge, student interest). This could include claims of "cultural neutrality" and appeals to a universal nature of mathematics.	There is at least one instance of connecting a math lesson to community/cultural knowledge and experience, such as during the lesson launch. Lesson briefly draws on student knowledge and experience, but they are not central to the lesson. The focus is with one student or a small group of students.	There is at least one sustained episode of sharing and developing collective understanding about mathematics that involves connecting to community/cultural knowledge to analyze authentic situations or issues in students' lives.	There are many sustained episodes of sharing and developing collective understandings about mathematics that involve connecting to cultural/community knowledge (e.g., student experiences are mathematized, student/parent connections with math work; math examples are embedded in local community/cultural contexts and activities such as games).	The creation and maintenance of collective understandings about mathematics that involves intricate connections to community/cultural knowledge permeates the entire lesson. This would include hook/intro, main activities, assessment, closure, and homework. Students are asked to analyze the mathematics within the community context and how the mathematics helps them understand that context.

STRAND: KNOWLEDGES AND IDENTITIES

2. **(Re)Humanizing Mathematics:** *How does my lesson support creativity, broaden what counts as mathematical knowledge, and affirm positive mathematical identities for all students?*

Fragile/
Margin →→→ Strong/
Centered

1	2	3	4	5
There is no evidence of humanizing practices. This could include mathematical knowledge treated as impersonal and unquestionable, mathematics of only the dominant school culture in the United States, and lack of connection to students as human beings.	There is some evidence of at least one aspect of humanizing practice in part or all of the lesson, which could include incorporating cultures and histories of students in the classroom, support for physical and emotional components of mathematical knowing, and students taking ownership of ideas or being asked to analyze/question mathematics as presented.	There are some instances of shared and collective construction of knowledge that • expands traditional notions of who can be good at mathematics, • may honor students' histories and cultures, • in other ways affirm mathematics identities across student groups, or being asked to analyze/question mathematics as presented.	There are many instances of shared and collective construction of knowledge that expands and challenges traditional notions of who can be good at mathematics and honors students' and/or marginalized people's histories, cultures, and perspectives *in service of* affirming mathematics identities.	There is a deliberate and continuous presence of humanizing practices, such as students drawing on many different knowledge bases to contribute to the construction of mathematical ideas, honoring of students' histories and different ways of knowing, in particular students from marginalized communities, as well as other forms of affirmation of mathematics identities.

STRAND: KNOWLEDGES AND IDENTITIES

3. Honoring Student Thinking & Ideas: *How does my lesson create opportunities to elicit, express, and build on student mathematical thinking in multiple ways?* (e.g., through gesture, pictures, words)

Fragile/Margin				Strong/Centered
1	**2**	**3**	**4**	**5**
The lesson does not include attention to student thinking.	The lesson includes some attention to student thinking.	The lesson includes at least two strategies aimed at making student thinking public.	The lesson includes multiple strategies to make student thinking public.	The lesson includes multiple strategies to make student thinking public.
Mathematical contributions in the lesson are almost exclusively from the teacher.	Teacher elicits student thinking of an individual student or small subset of students.	Teacher elicits student thinking among students in at least one phase of the lesson (launch, explore, or summary).	Teacher elicits mathematical thinking across all phases of the lesson.	Teacher and students elicit mathematical thinking across all phases of the lesson.
Shared understanding or collective meaning making is absent.	Sharing of mathematical ideas is among a few select students or between a student and the teacher.	Shared understanding about mathematical ideas and contributions are evident in at least one part of the lesson.	Multiple forms of student mathematical contributions are encouraged and valued by teacher and students.	All contributions are valued and respected by teachers and students.
	Shared understanding is minimal.		Shared understanding between teacher and students as well as among students is evident across the lesson.	There are multiple and sustained opportunities for teachers and students to collectively respond to each other's thinking and contribute to refining mathematical ideas core to student learning.

STRAND: RIGOR AND SUPPORT

4. **Sustaining High Cognitive Demand:** *How does my lesson enable all my students to closely explore and analyze math concepts(s), procedure(s), and problem-solving/reasoning strategies?*

Fragile/ Margin		Strong/ Centered	
1	**2**	**3**	**4**	**5**

1	**2**	**3**	**4**	**5**
Students receive, recite, or memorize facts, procedures, and definitions. There is no evidence of conceptual understanding being required. There are no opportunities for mathematical problem-solving, mathematical analysis, or exploration.	Students primarily receive, recite, or perform routine procedures without analysis or connection to underlying concepts or mathematical structure. There are some opportunities for mathematical exploration, but activities do not require analysis to complete. OR A select group of students get access to activities requiring authentic problem-solving, analysis of procedures, concepts, or underlying mathematical structure.	At least one sustained activity involves all students with complex problem-solving, analysis of procedures, concepts, or underlying mathematical structure. There is at least one sustained activity that requires mathematical exploration, analysis, and explanation.	Most of the lesson involves all students in activities that require close analysis of procedures, concepts, or underlying mathematical structure. OR involve complex mathematical thinking, use multiple representations, and demand justification.	The entire lesson involves all students in activities that require close analysis of procedures and concepts, involve complex mathematical thinking, use multiple representations, AND demand explanation and justification.

STRAND: RIGOR AND SUPPORT

5. Scaffolding Up: *How does my lesson maintain high rigor with high support for all students?*

Fragile/ Margin				Strong/ Centered
1	2	3	4	5
There is no evidence that the teacher has planned supports in ways that maintain the rigor of the task while providing access for students.	Planned supports provide too much scaffolding and diminish the rigor of the task. OR Planned supports may only attend to access at the start of the task, not throughout the lesson.	Planned supports maintain rigor but may not connect to either this *specific mathematics task*, or draw on the strengths of students. Planned supports may only attend to access at the start of the task, not throughout the lesson.	Specific planned supports ensure most of the class understands the task and has a way to get started. Planned supports are used throughout the lesson, although planned supports for individuals or subgroups may not directly connect to known student strengths. There is no evidence or minimal evidence of supports planned or used for individuals or subgroups of students.	Specific, planned supports address the whole class, as well as individual or subgroup needs. Planned supports are used throughout all phases of the lesson, including launch, students, work time, strategy sharing, or lesson wrap up. Planned and enacted supports for subgroups differ from those for the whole class and build from students' known strengths.

STRAND: RIGOR AND SUPPORT

6. Affirming Multilingualism: *How does my lesson make space for multilingual learners (MLL) to be central participants in mathematics activities?*

Fragile/ Margin				Strong/ Centered
1	**2**	**3**	**4**	**5**
There is no acknowledgment of MLLs' linguistic funds of knowledge.	There is acknowledgment of MLLs' linguistic funds of knowledge, but they are not leveraged in lesson design. Students' use of L1 is tolerated.	There is at least one instance of attention to MLLs' linguistic funds of knowledge that is central to the lesson, such as encouraging translanguaging.	Clear attention is paid to MLLs' linguistic funds of knowledge throughout the lesson.	Extensive and sustained attention is paid to MLLs' linguistic funds of knowledge throughout lesson.
MLLs who are not yet fully proficient in English are ignored and/or seated apart from their classmates.	Teaching focuses on correct usage of English vocabulary only.	Even if a teacher does not use L1, it is evident that MLLs' linguistic repertoires are valued and that they are encouraged to build on them (e.g., students can present in L1, students work in groups in L1).	Focus is on mathematical discourse in L1 and English, not students' production of "correct" English.	Sustained encouragement of L1 usage, or hybrid language (e.g., code-switching) is observed between teacher and students and among students, in a variety of interactions (teacher-students, pair, small group, and whole class). The main focus is the development of mathematical discourse and meaning making in both L1 and English.
	There is no explicit attention to scaffolding access for MLLs.	There is at least one instance in which an English as a Second Language (ESL) scaffolding strategy is used to develop academic language (i.e., revoicing, use of graphic organizers, activation of prior knowledge, strategic grouping with bilingual students).	There is sustained use of at least two ESL scaffolding strategies, such as the use of revoicing and attention to cognates, direct modeling of vocabulary, strategic grouping with bilingual students, use of realia, graphic organizers, or encouragement of L1 usage is observed at least between teacher and one student or small group of students.	Deliberate and continuous use of multiple ESL strategies, such as gesturing, use of realia, use of cognates, revoicing, graphic organizers and manipulatives are observed during whole class, and /or small group instruction and discussions. The main focus is the development of mathematical discourse, identity, and meaning making as learners are positioned as mathematically competent leaders and thinkers.
			The focus is on positioning of multilingual students as central participants through recognizing their mathematical competence.	

STRAND: POWER AND PARTICIPATION

7. Distributing Intellectual Authority: *How does my lesson distribute mathematics authority and make space for multiple forms of knowledge and communication?*

Fragile/
Margin   →   Strong/
Centered

1	2	3	4	5
The authority of math knowledge exclusively resides with the teacher (e.g., tightly controls talk in the classroom, teacher decides what answer is correct, IRE patterns may be evident in classroom discourse). Student participation is severely limited (e.g., limited to one-word answers, short choral responses, repetition of teacher).	The authority of mathematics knowledge is infrequently shared and primarily resides with the teacher and a few students. Student participation is limited (e.g., limited to one-word answers, short choral responses, repetition of teacher).	The authority of math knowledge between teacher and students is sporadically shared and resides with teacher and some students. Some students participate in math activities in substantive ways, periodically sharing reasoning and different strategies, and understanding the strategies of others.	The authority of math knowledge is equally shared among teacher and many students. Most students participate in mathematical activity in substantive ways, and frequently communicate mathematical ideas in at least two modalities (e.g., listening, writing, drawing, speaking, gestures).	The authority of math knowledge is widely shared among teacher and most students, and *students* hold most of the math authority. All students participate in mathematical activities in substantive ways and communicate mathematical ideas through multiple modalities (e.g., listening, writing, drawing, speaking, gestures).

STRAND: POWER AND PARTICIPATION

8. **Disrupting Status and Power:** *How does my lesson disrupt status differences, entrenched stereotypes, and inequitable power relationships present in all mathematics classrooms?*

Fragile/Margin				Strong/Centered
1	**2**	**3**	**4**	**5**
No strategies to minimize status issues are evident. Student involvement is structured to privilege a dominant subgroup (in terms of race, class, gender, language, (dis)ability, and other socially constructed identities).	At least one strategy to minimize status differences is evident but superficial and does not challenge stereotypes or other power dynamics. Student involvement is structured to privilege a dominant subgroup (in terms of race, class, gender, language, (dis)ability, and other markers of status, with limited involvement from nondominant students.	Some strategies to minimize status differences among students (and specific subgroups) in the lesson are evident and have some effect. Strategies may have a momentary impact on some subgroup but may not necessarily address a persistent status issue related to race, gender, (dis)ability, language, and other markers of privilege. Student involvement is structured to support particular subgroups, which may include some but not all nondominant groups.	Some strategies to minimize status differences among students (and specific subgroups) are evident and have some effect. Teacher uses one or more strategies that • maximize student mathematical, cultural, and linguistic strengths, • explicitly address stereotypes, and • structure compassionate and inclusive talk (e.g., building each other up, not tearing down) Student involvement is structured to support most nondominant subgroups.	Multiple strategies to minimize status differences among students (and specific subgroups) are implemented effectively throughout the lesson. Teacher and students both work to minimize status issues through strategies that • maximize student mathematical, cultural, and linguistic strengths. • explicitly address stereotypes, and • structure compassionate and inclusive talk (e.g., building each other up, not tearing down). Student involvement is structured to support multiple or all subgroups, with particular attention to historically marginalized and segregated students.

STRAND: POWER AND PARTICIPATION

9. Analyzing and Taking Action: *How does my lesson support student use of mathematics to analyze, critique, and address power relationships and injustice in their lives (economic, social, environmental, legal, political, patriarchal)?*

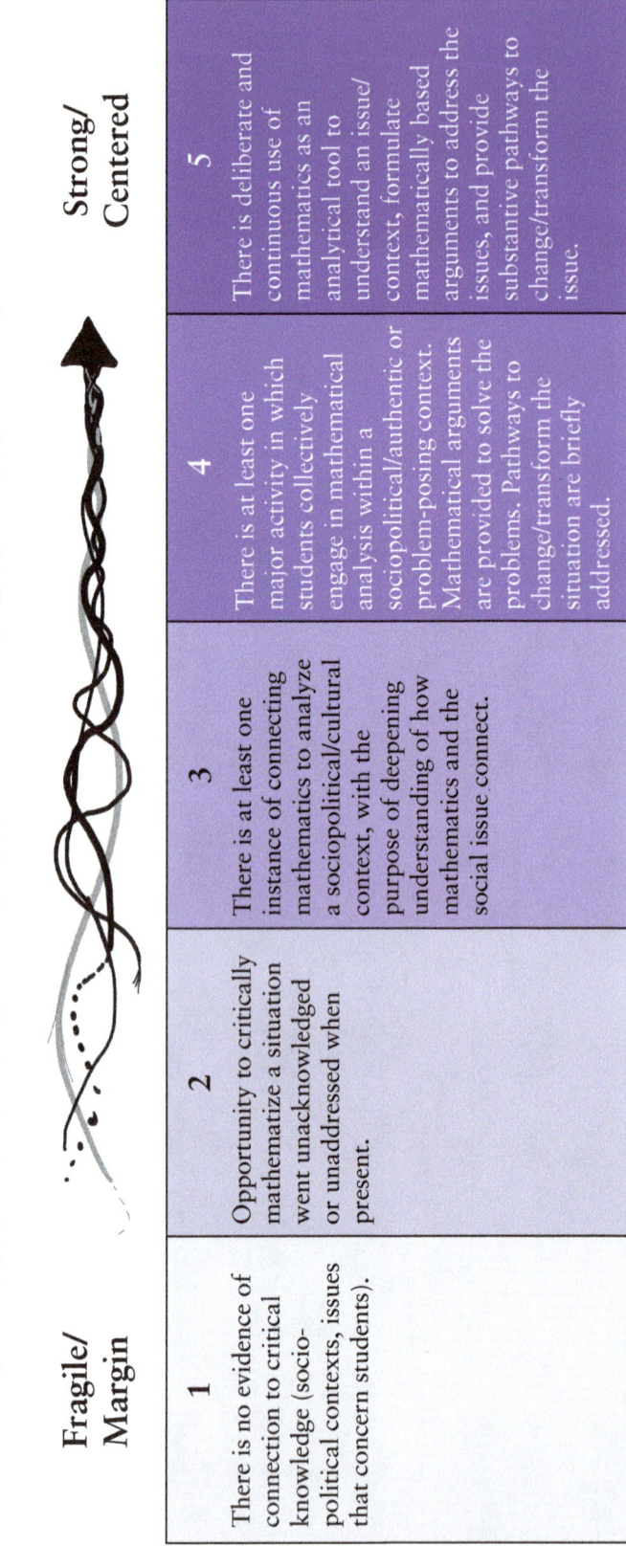

Fragile/ Margin				Strong/ Centered
1	**2**	**3**	**4**	**5**
There is no evidence of connection to critical knowledge (socio-political contexts, issues that concern students).	Opportunity to critically mathematize a situation went unacknowledged or unaddressed when present.	There is at least one instance of connecting mathematics to analyze a sociopolitical/cultural context, with the purpose of deepening understanding of how mathematics and the social issue connect.	There is at least one major activity in which students collectively engage in mathematical analysis within a sociopolitical/authentic or problem-posing context. Mathematical arguments are provided to solve the problems. Pathways to change/transform the situation are briefly addressed.	There is deliberate and continuous use of mathematics as an analytical tool to understand an issue/ context, formulate mathematically based arguments to address the issues, and provide substantive pathways to change/transform the issue.

SOURCE: Adapted from Aguirre et al. (2015); Aguirre & Zavala (2013); CEMELA (2006); Kitchen (2005); Turner et al. (2012).

REFERENCES

Adams, M. (2018). "I can solve all the problems": Latinx students (re)write their math stories. In I. Goffney & R. Gutiérrez (Eds.), *Rehumanizing mathematics for students who are Black, Indigenous, and/or Latinx students* (pp. 121–134). Annual perspectives in mathematics education. National Council of Teachers of Mathematics.

Aguirre, J. M. (2009). Privileging mathematics and equity in teacher education: Framework, counter-resistance strategies and reflections from a Latina mathematics educator. In B. Greer, S. Mukhopadhyay, S. Nelson-Barber, & A. Powell (Eds.) *Culturally responsive mathematics education* (pp. 295–319). Routledge.

Aguirre, J. M., & Civil, M. (Eds.). (2016). Special issue on mathematics education through the lens of social justice. *Teaching for Excellence and Equity in Mathematics*, 7(1).

Aguirre, J. M., Foote, M. Q., Turner, E. E., Bartell, T. G., Drake, C., & Roth McDuffie, A. (2015, November). Supporting new K–8 teachers to be culturally responsive using a lesson analysis tool. In T. G. Bartell, K. N. Bieda, R. T. Putnam, K. Bradfield, & H. Dominguez (Eds.) *Proceedings of the 37th Annual Meeting of the North American Chapter of the International Group for the Psychology of Mathematics Education* (pp. 868–871). Michigan State University.

Aguirre, J. M., Mayfield-Ingram, K., & Martin, D. B. (2013). *The impact of identity in K–8 mathematics learning and teaching: Rethinking equity-based practices*. National Council of Teachers of Mathematics.

Aguirre, J. M., Suh, J., Tate, H., Carlson, M. A., Fulton, E. A., & Turner, E. E. (2022). Leveraging equity and civic empathy through community-based mathematical modeling. In A. Lischka, E. Dyer, R. Jones, J. Lovett, J. Strayer, & S. Drown (Eds.), *Proceedings of the Forty-fourth Annual Meeting of the North American Chapter of the International Group for the Psychology of Mathematics Education* (pp. 349–358). Middle Tennessee State University.

Aguirre, J. M., Turner, E. E., McVicar, E., Roth McDuffie, A., Foote, M. Q., & Carl, E. (2022). *Mathematizing the world routine: Leveraging children's multiple mathematical knowledge bases in the elementary classroom*. Manuscript submitted for publication.

Aguirre, J. M., & Zavala, M. (2013). Making culturally responsive mathematics teaching explicit: A lesson analysis tool. *Pedagogies: An International Journal*, 8(2), 163–190.

Aguirre, J. M., Zavala, M. D. R., & Katanyoutanant, T. (2012). Developing robust forms of pre-service teachers' pedagogical content knowledge through culturally responsive mathematics teaching analysis. *Mathematics Teacher Education and Development*, 14(2), 113–136.

Anghileri, J. (2006). Scaffolding practices that enhance mathematics learning. *Journal of Mathematics Teacher Education*, 9, 33–52.

Anhalt, C. O. (2014). Scaffolding in mathematical modeling for ELLs. In E. Turner & M. Civil (Eds.), *The common core state standards in mathematics for English language learners: Grades K–8* (pp. 111–126). Teaching English to Speakers of Other Languages (TESOL) International Association.

Arnold, E. G., Burroughs, E. A., Carlson, M. A., Fulton, E. W., & Wickstrom, M. H. (2021). *Becoming a teacher of mathematical modeling: Grade K–5*. National Council of Teachers of Mathematics.

Barajas-López, F., & Bang, M. (2018). Indigenous making and sharing: Claywork in an Indigenous STEAM program. *Equity & Excellence in Education, 51*(1), 7–20.

Behrend, J. (2001). Are the rules interfering with children's mathematical understanding? *Teaching Children Mathematics, 8*(1), 36–40.

Berry, R. Q., III, Ellis, M., & Hughes, S. (2014). Examining a history of failed reforms and recent stories of success: Mathematics education and Black learners of mathematics in the United States. *Race Ethnicity and Education, 17*(4), 540–568.

Boaler, J. (2002). *Experiencing school mathematics: Traditional and reform approaches to teaching and their impact on student learning.* Routledge.

Boaler, J. (2015). *Mathematical mindsets: Unleashing students' potential through creative math, inspiring messages and innovative teaching.* John Wiley & Sons.

Boaler, J., & Selling, S. K. (2017). Psychological imprisonment or intellectual freedom? A longitudinal study of contrasting school mathematics approaches and their impact on adults' lives. *Journal for Research in Mathematics Education, 48*(1), 78–105. https://doi.org/10.5951/jresematheduc.48.1.0078

Boston, M. D., & Smith, M. S. (2009). Transforming secondary mathematics teaching: Increasing the cognitive demands of instructional tasks used in teachers' classrooms. *Journal for Research in Mathematics Education, 40*(2), 119–156.

Cabrera, N. L., Milem, J. F., Jaquette, O., & Marx, R. W. (2014). Missing the (student achievement) forest for all the (political) trees: Empiricism and the Mexican American studies controversy in Tucson. *American Educational Research Journal, 51*(6), 1084–1118.

Carpenter, T. P., Fennema, E., Franke, M. L., Levi, L., & Empson, S. B. (1999/2014). *Children's mathematics: Cognitively guided instruction.* Heinemann.

Celedón-Pattichis, S., & Ramirez, N. G. (2012). *Beyond good teaching: Advancing mathematics education for ELLs.* National Council of Teachers of Mathematics.

Center for Mathematics Education of Latinos/as. (CEMELA). 2006. *Mathematics classroom observation protocol.* University of Arizona.

Chao, T., & Jones, D. (2016). That's not fair and why: Developing social justice mathematics activists in pre-K. *Teaching for Excellence and Equity in Mathematics, 7*(1), 15–21.

Children's Equity Project. (2022). *A holistic approach to ending exclusionary discipline for young learners: A review of the data, research, and multidimensional solutions.* https://childandfamilysuccess.asu.edu/sites/default/files/2022-09/exclusionary-discipline-093022-1.pdf

Chval, K. B., Smith, E., Trigos-Carrillo, L., & Pinnow, R. J. (2021). *Teaching math to multilingual students: Positioning English learners for success.* Corwin.

Civil, M. (2007). Building on community knowledge: An avenue to equity in mathematics education. In N. S. Nasir & P. Cobb (Eds.), *Improving access to mathematics: Diversity and equity in the classroom* (pp. 105–117).

Civil, M., & Khan, L. H. (2001). Mathematics instruction developed from a garden theme. *Teaching Children Mathematics, 7*(7), 400–405.

Civil, M., & Menéndez, J. M. (2010). *Involving Latino parents in their children's mathematics education.* NCTM Research Brief.

Civil, M., & Turner, E. (2014). *The common core state standards in mathematics for English language learners: Grades K–8.* Tesol Press.

Cohen, E. G., & Lotan, R. A. (1995/2014). *Designing groupwork: Strategies for the heterogeneous classroom (3rd ed.).* Teachers College Press.

Crenshaw, K. W., Ocen, P., & Nanda, J. (2015). *Black girls matter: Pushed out, overpoliced and underprotected.* African American Policy Forum.

de Araujo, Z., Roberts, S. A., Willey, C., & Zahner, W. (2018). English learners in K–12 mathematics education: A review of the literature. *Review of Educational Research, 88*(6), 879–919.

Dominguez, H., LopezLeiva, C., & Khisty, L. L. (2014). Relational engagement: Proportional reasoning with bilingual Latino/a students. *Educational Studies in Mathematics, 85,* 143–160.

Dunleavy, T. K. (2015). Delegating mathematical authority as a means to strive toward equity. *Journal of Urban Mathematics Education, 8*(1), 62–82.

Dweck, C. S. (2006). *Mindset: The new psychology of success.* Random House.

Eglash, R. (1999). *African fractals: Modern computing and indigenous design.* Rutgers University Press.

Ellis, M. (2008). Leaving no child behind yet allowing none too far ahead: Ensuring (in)equity in mathematics education through the science of measurement and instruction. *Teachers College Record, 110*(6), 1330–1356.

Empson, S., Levi, L., & Carpenter, T. P. (2011). The algebraic nature of fractions: Developing relational thinking in elementary school. In J. Cai & E. Knuth (Eds.), *Early algebraization: Advances in mathematics education* (pp. 409–429). Springer.

EQ-STEMM. (2022). *Teacher moves table: Advancing equity and strengthening teaching with elementary mathematics modeling.* www.eqstemm.org

Featherstone, H., Crespo, S., Jilk, L. M., Oslund, J. A., Parks, A. N., & Wood, M. B. (2011). *Smarter together! Collaboration and equity in the elementary math classroom.* National Council of Teachers of Mathematics.

Flores, A. (2007). Examining disparities in mathematics education: Achievement gap or opportunity gap? *The High School Journal, 91*(1), 29–42.

Frankenstein, M. (1983). Critical mathematics education: An application of Paulo Freire's epistemology. *Journal of Education, 165*(4), 315–339.

Freire, P. (1970/1993). *Pedagogy of the oppressed.* Continuum.

Furuto, L. (2014). Pacific ethnomathematics: Pedagogy and practices in mathematics education. *Teaching Mathematics and Its Applications: An International Journal of the IMA, 33*(2), 110–121.

Furuto, L. (2015). *Hōkūle'a worldwide voyage: Island wisdom and global connections in ethnomathematics.* University of Hawaii. https://scholarspace.manoa.hawaii.edu/items/d2fbdec7-7a41-4563-8d0e-87e890771142

García, O. (2009). *Bilingual education in the 21st century: A global perspective.* Wiley-Blackwell.

García, O., Johnson, S. I., Seltzer, K., & Valdés, G. (2017). *The translanguaging classroom: Leveraging student bilingualism for learning.* Caslon.

Garfunkel, S., & Montgomery, M. (2019). *GAIMME: Guidelines for assessment and instruction in mathematical modeling education* (2nd ed.). A joint publication of the Consortium for Mathematics and Its Applications & Society for Industrial and Applied Mathematics.

Gay, G. (2000). *Culturally responsive teaching: Theory, research, and practice.* Teachers College Press.

Gerdes, P. (1985). Conditions and strategies for emancipatory mathematics education in undeveloped countries. *For the Learning of Mathematics, 5*(1), 15–20.

Gewertz, C. (2019). Seattle schools lead controversial push to "rehumanize" math: Classes would explore power and oppression. *Education Week.* https://www.edweek.org/teaching-learning/seattle-schools-lead-controversial-push-to-rehumanize-math/2019/10

Gholson, M. L. (2016). Clean corners and algebra: A critical examination of the constructed invisibility of Black girls and women in mathematics. *Journal of Negro Education, 85*(3), 290–301.

Gholson, M. L., & Robinson, D. D. (2019). Restoring mathematics identities of Black learners: A curricular approach. *Theory Into Practice, 58*(4), 347–358.

Gilmer, G. (n.d.). *Mathematical patterns in African American hairstyles.* http://www.math.buffalo.edu/mad/special/gilmer-gloria_HAIRSTYLES.html

Goffney, I., Gutiérrez, R., & Boston, M. (Eds.). (2018). *Rehumanizing mathematics for Black, Indigenous, and Latinx students.* Annual perspectives in mathematics education. National Council of Teachers of Mathematics.

Gonzalez, N., Moll, L. C., & Amanti, C. (2005). *Funds of knowledge.* Routledge.

Gottbrath, L. (2020). *In 2020, the Black Lives Matter movement shook the world* [Photograph]. Aljazeera. https://www.aljazeera.com/features/2020/12/31/2020-the-year-black-lives-matter-shook-the-world

Grossman, P., Schoenfeld, A., & Lee, C. (2005). Teaching subject matter. In L.

Darling-Hammond & J. Bransford (Eds.), *Preparing teachers for a changing world* (pp. 201–231). Jossey-Bass.

Gutiérrez, R. (2008). Research commentary: A gap-gazing fetish in mathematics education? Problematizing research on the achievement gap. *Journal for Research in Mathematics Education, 39*(4), 357–364.

Gutiérrez, R. (2010/2013). The sociopolitical turn in mathematics education. *Journal for Research in Mathematics Education, 44*(1), 37–68.

Gutiérrez, R. (2013). Why (urban) mathematics teachers need political knowledge. *Journal of Urban Mathematics Education, 6*(2), 7–19.

Gutiérrez, R. (2015). Nesting in Nepantla: The importance of maintaining tensions in our work. In N. Joseph, C. Haynes, & F. Cobb (Eds.), *Interrogating whiteness and relinquishing power: White faculty's commitment to racial consciousness in STEM classrooms* (pp. 253–282). Peter Lang.

Gutiérrez, R. (2016). Strategies for creative insubordination in mathematics teaching. *Special Issue Mathematics Education: Through the Lens of Social Justice, 7*(1), 52–62.

Gutiérrez, R. (2018). The need to rehumanize mathematics. In I. Goffney & R. Gutiérrez (Eds.), *Rehumanizing mathematics for Black, Indigenous and Latinx students*. Annual perspectives in mathematics education (pp. 1–10). National Council of Teachers of Mathematics.

Gutstein, E. (2006). *Reading and writing the world with mathematics: Toward a pedagogy for social justice.* Taylor & Francis.

Gutstein, E., & Petersen, B. (2013). *Rethinking mathematics: Teaching social justice by the numbers* (2nd ed.). Rethinking Schools.

Hampson, R. (2013). *What you didn't know about King's "Dream" speech* [Photograph]. USA Today. https://www.usatoday.com/story/news/nation/2013/08/12/march-on-washington-king-speech/2641841/

Henner, J., Pagliaro, C., Sullivan, S., & Hoffmeister, R. (2021). Counting differently. *American Annals of the Deaf, 166*(3), 318–341.

Horn, I. S. (2007). Fast kids, slow kids, lazy kids: Framing the mismatch problem in mathematics teachers' conversations. *The Journal of the Learning Sciences, 16*(1), 37–79.

Horn, I. S. (2012). *Strength in numbers: Collaborative learning in secondary classrooms.* National Council of Teachers of Mathematics.

Indiana University. (2023). *Latino studies program* [Photograph]. https://latinostudies.indiana.edu/graduate/courses.html

Jones, S. R., & Marchant, C. N. G. (2022). Twin skin of Raza learners: Race, language, and mathematics. *Mathematics Teacher: Learning and Teaching PK–12, 115*(1), 45–48.

Joseph, G. G. (1991/2011). *The crest of the peacock: Non-European roots of mathematics.* Princeton University Press.

Joseph, N. (2022). *Making Black girls count in mathematics education: A Black feminist vision for transforming teaching.* Harvard Education Press.

Kirchner, M. K., & Sarhangi, R. (2011). Connecting the art of Navajo weavings to secondary education. *Ohio Journal of School Mathematics, 64*, 11–17.

Kitchen, R. S. (2005). Making equity and multiculturalism explicit to transform mathematics education. In A. J. Rodriguez & R. Kitchen (Eds), *Preparing mathematics and science teachers for diverse classrooms: Promising strategies for transformative pedagogy* (pp. 33–60). Lawrence Erlbaum Associates.

Kitchen, R. S., DePree, J., Celed, S., & Brinkerhoff, J. (2007). *Mathematics education at highly effective schools that serve the poor: Strategies for change.* Routledge.

Ladson-Billings, G. (1995). Toward a theory of culturally relevant pedagogy. *American Educational Research Journal, 32*(3), 465–491.

Lambert, R. (2018). "Indefensible, illogical, and unsupported"; Countering deficit mythologies about the potential of students with learning disabilities in mathematics. *Education Sciences, 8*(2), 72. https://doi.org/10.3390/educsci8020072

Lampert, M. (2001). *Teaching problems and the problems of teaching.* Yale University Press.

Langer-Osuna, J. M. (2018). Exploring the central role of student authority relations in collaborative mathematics. *ZDM, 50*(6), 1077–1087.

Learning for Justice. (2022). *Social justice standards: The learning for justice anti-bias framework*. The Southern Poverty Law Center.

Looney Math Consulting. (2022). *Same but different math*. https://www.samebutdifferent math.com/

Macedo, D., Dendrinos, B., & Gounari, P. (2003). *The hegemony of English*. Routledge.

Maldonado Rodriguez, L. A., Krause, G., & Adams-Corral, M. (2020). Flowing with the translanguaging corriente: Juntos engaging with and making sense of mathematics. *Teaching for Excellence and Equity in Mathematics, 11*(2), 17–25.

Martin, D. B. (2000). *Mathematics success and failure among African-American youth: The roles of sociohistorical context, community forces, school influence, and individual agency*. Routledge.

Martin, D. B. (2012). Learning mathematics while Black. *Educational Foundations, 26*, 47–66.

Martin, D. B. (2015). The collective Black and principles to actions. *Journal of Urban Mathematics Education, 8*(1), 17–23.

Martin, D. B., Gholson, M. L., & Leonard, J. (2010). Mathematics as gatekeeper: Power and privilege in the production of knowledge. *Journal of Urban Mathematics Education, 3*(2), 12–24.

Mattox, N. (2021). The Teacher Space. In N. Mattox, *Black calculus*. Nomadic Press.

McGee, E. O. (2020). Interrogating structural racism in STEM higher education. *Educational Researcher, 49*(9), 633–644.

Mirra, N. (2018). *Educating for empathy: Literacy learning and civic engagement*. Teachers College Press.

Moschkovich, J. (1999). Supporting the participation of English language learners in mathematical discussions. *For the Learning of Mathematics, 19*(1), 11–19.

Moschkovich, M. (2013). Principles and guidelines for equitable mathematics teaching practices and materials for English Language Learners. In Stinson, D. W., & Spencer, J. A. (Eds.). Privilege and oppression in the mathematics preparation of teacher educators [Special issue]. *Journal of Urban Mathematics Education, 6*(1), pp. 45–57.

Muhammad, G. (2023). *Unearthing joy: A guide to culturally and historically responsive curriculum and instruction*. Scholastic Incorporated.

National Governor's Association Center for Best Practices & Council of Chief State School Officers. (2010). *Common core state standards for mathematics*. Authors.

National Research Council. (2001). *Adding it up: Helping children learn mathematics*. National Academy Press.

Ni, Y., Zhou, D. H. R., Cai, J., Li, X., Li, Q., & Sun, I. X. (2018). Improving cognitive and affective learning outcomes of students through mathematics instructional tasks of high cognitive demand. *The Journal of Educational Research, 111*(6), 704–719.

Oakes, J. (1985/2005). *Keeping track: How schools structure inequality*. Yale University Press.

Perkins, I., & Flores, A. (2002). Mathematical notations and procedures of recent immigrant students. *Mathematics Teaching in the Middle School, 7*(6), 346–351.

Pinxten, R. (1997). Applications in the teaching of mathematics and sciences. In *Ethnomathematics: Challenging Eurocentrism in mathematics education* (pp. 373–401).

Powell, A. B., & Frankenstein, M. (Eds.). (1997). *Ethnomathematics: Challenging Eurocentrism in mathematics education*. State University of New York Press.

Reuters. (2022). *How Sri Lankan protests unfolded* [Photograph]. https://www.reuters.com/world/asia-pacific/how-sri-lankan-protests-unfolded-2022-07-09/

Reynolds, P. H. (2019). *Say something*. Scholastic.

Robinson, E. (2022). *Protestors rally in solidarity with Iranian people* [Photograph]. Spectrum News NY1. https://www.ny1.com/nyc/manhattan/news/2022/10/15/protesters-rally-in-solidarity-with-iranian-people

Roth, W. M., & Radford, L. (2010). Re/thinking the zone of proximal development (symmetrically). *Mind, Culture, and Activity, 17*(4), 299–307.

Rubel, L. H., Lim, V., Hall-Wieckert, M., & Katz, S. (2016). Cash across the city: Participatory mapping & teaching for spatial justice.

Journal of Urban Learning, Teaching, & Research, 12, 4–14.

Rubel, L. H., & Nicol, C. (2020). The power of place: Spatializing critical mathematics education. *Mathematical Thinking and Learning, 22*(3), 173–194.

Seda, P., & Brown, K. (2021). *Choosing to see: A framework for equity in the math classroom.* Dave Burgess Consulting.

Shagoury, R., & Power, B. M. (2012). *Living the questions: A guide for teacher-researchers.* Stenhouse.

Simic-Muller, K., Turner, E. E., & Varley, M. C. (2009). Math club problem posing. *Teaching Children Mathematics, 16*(4), 206–212.

Smith, M. S., & Stein, M. K. (1998). Reflections on practice: Selecting and creating mathematical tasks: From research to practice. *Mathematics Teaching in the Middle School, 3*(5), 344–350.

Smith, M. S., & Stein, M. (2018). *5 practices for orchestrating productive mathematics discussions* (2nd ed.). Corwin.

Sowder, J. T. (2007). The mathematics education and development of teachers. In F. K. Lester (Ed.), *Second handbook of research on mathematics teaching and learning* (pp. 157–223). Information Age.

Steele, J. L., Slater, R. O., Zamarro, G., Miller, T., Li, J., Burkhauser, S., & Bacon, M. (2017). Effects of dual-language immersion programs on student achievement: Evidence from lottery data. *American Educational Research Journal, 54*(1_suppl), 282S–306S. https://doi.org/10.3102/0002831216634463

Stein, M. K., Smith, M. S., Henningsen, M., & Silver, E. A. (2000). *Implementing standards-based mathematics instruction: A casebook for professional development.* Teachers College Press.

Stein, M. K., Smith, M. S., & Silver, E. (1999). The development of professional developers: Learning to assist teachers in new settings in new ways. *Harvard Educational Review, 69*(3), 237–270.

Suh, J. M., Matson, K., & Seshaiyer, P. (2017). Engaging elementary students in the creative process of mathematizing their world through mathematical modeling. *Education Sciences, 7*(62), 1–21.

Suh, J., Matson, K., Seshaiyer, P., Jamieson, S., & Tate, H. (2021). Mathematical modeling as a catalyst for equitable mathematics instruction: Preparing teachers and young learners with 21st century skills. *Mathematics, 9*(2), 162.

Suh, J. M., Wickstrom, M. H., & English, L. D. (Eds.). (2021). *Exploring mathematical modeling with young learners.* Springer International.

Tan, P., Padilla, A., & Lambert, R. (2022). A critical review of educator and disability research in mathematics education: A decade of dehumanizing waves and humanizing wakes. *Review of Educational Research, 92*(6), 871–910. https://doi.org/10.3102/00346543221081874

Tate, H., Proffitt, T., Christensen, A., Hunter, C., Stratton, D., Fleshman, E., Aguirre, J., & Suh, J. (2022). Mathematizing representation in children's libraries: An anti-racist math unit in elementary grades. *Teaching for Excellence and Equity in Mathematics—Special Issue: Anti-racism in Mathematics Education. TODOS: Mathematics for All, 13*(1), 23–40.

Tate, W. (1994). Race, retrenchment, and the reform of school mathematics. *The Phi Delta Kappan, 75*(6), 477–484.

TODOS: Mathematics for All. (2020). *The mo(ve)ment to prioritize antiracist mathematics: Planning for this and every school year.* https://bit.ly/3j5Yvip

Tucker-Raymond, E., Varelas, M., Pappas, C. C., Korzh, A., & Wentland, A. (2007). "They probably aren't named Rachel": Young children's scientist identities as emergent multimodal narratives. *Cultural Studies of Science Education, 1,* 559–592.

Turner, E. E., Aguirre, J. M., Foote, M. Q., Anhalt, C. O., & Roth McDuffie, A. (2018, June). *Mathematizing the world: Routines and tasks that foster mathematical modeling with cultural and community contexts* [Conference presentation]. TODOS: Mathematics for All.

Turner, E. E., Aguirre, J. M., Roth McDuffie, A., & Foote, M. Q. (2019, November). *Jumping into modeling: Elementary math modeling with school and community contexts* [Research presentation]. Annual North American Chapter meeting of the Psychology of Mathematics Education, East Lansing, MI.

Turner, E. E., Bennett, A. B., Granillo, M., Ponnuru, N., Roth Mcduffie, A., Foote, M. Q., Aguirre, J. M., & McVicar, E. (2022). Authenticity of elementary teacher designed and implemented mathematical modeling tasks. *Mathematical Thinking and Learning*, 1–24. https://doi.org/10.1080/10986065.2022.2028225

Turner, E. E., & Celedón-Pattichis, S. (2011). Mathematical problem solving among Latina/o kindergartners: An analysis of opportunities to learn. *Journal of Latinos and Education*, 10(2), 146–169.

Turner, E. E., Drake, C., McDuffie, A. R., Aguirre, J., Bartell, T. G., & Foote, M. Q. (2012). Promoting equity in mathematics teacher preparation: A framework for advancing teacher learning of children's multiple mathematics knowledge bases. *Journal of Mathematics Teacher Education*, 15, 67–82. https://doi.org/10.1007/s10857-011-9196-6

Turner, E. E., Roth McDuffie, A., Aguirre, J. M., Foote, M. Q., Chappelle, C., Bennett, A., Granillo, M., & Ponnuru, N. (2021). Upcycling plastic bags to make jump ropes: Elementary students leverage experiences and knowledge as they engage in a relevant community-oriented mathematical modeling task. In J. Suh, M. H. Wickstrom, & L. English (Eds.), *Exploring the nature of mathematical modeling in the early grades* (pp. 235–266). Springer.

Ullrich, J. S. (2019). For the love of our children: An indigenous connectedness framework. *AlterNative: An International Journal of Indigenous Peoples*, 15(2), 121–130. https://doi.org/10.1177/1177180119828114

Vygotsky, L. S. (1978). *Mind in society: The development of higher psychological processes*. Harvard University Press.

Vygotsky, L. (1986). *Though and language*. MIT.

Warren, B., & Rosebery, A. S. (1991). *Cheche Konnen: Collaborative scientific inquiry in language minority classrooms. A handbook for teachers and planners from the innovative approaches research project* (2nd ed.). Teacher Education Research Center.

Webb, N. M., Franke, M. L., Ing, M., Wong, J., Fernandez, C. H., Shin, N., & Turrou, A. C. (2014). Engaging with others' mathematical ideas: Interrelationships among student participation, teachers' instructional practices, and learning. *International Journal of Educational Research*, 63, 79–93.

Webel, C. (2010). Connecting research to teaching: Shifting mathematical authority from teacher to community. *The Mathematics Teacher*, 104(4), 315–318.

Wells, C. L. (2018). Understanding issues associated with tracking students in mathematics education. *Journal of Mathematics Education*, 11(2), 68–84.

Welsh, R. O., & Little, S. (2018). The school discipline dilemma: A comprehensive review of disparities and alternative approaches. *Review of Educational Research*, 88(5), 752–794.

Wyborney, S. (2023). The estimation clipboard [Image]. Steve Wyborney's Blog: I'm on a Learning Mission. https://stevewyborney.com/2018/04/the-estimation-clipboard/

Yeh, C., Ellis, M. W., & Koehn Hurtado, C. (2017). *Reimagining the mathematics classroom*. National Council of Teachers of Mathematics.

Yeh, C., & Otis, B. M. (2019). Mathematics for whom: Reframing and humanizing mathematics. *Occasional Paper Series*, (41). https://doi.org/10.58295/2375-3668.1276

Yosso, T. J. (2005). Whose culture has capital? A critical race theory discussion of community cultural wealth. *Race Ethnicity and Education*, 8(1), 69–91.

Zavala, M. (2017). Bilingual pre-service teachers grapple with the academic and social role of language in mathematics discussions. *Issues in Teacher Education*, 26(2), 49–66.

Zavala, M. (2023). Examining air quality. In C. Koestler, J. Ward, M. Zavala, & T. G. Bartell (Eds.). *Early elementary mathematics lessons to explore, understand, and respond to social injustice*. Corwin & NCTM.

Zavala, M. D. R., & Hand, V. (2019). Conflicting narratives of success in mathematics and science education: Challenging the achievement-motivation master narrative. *Race Ethnicity and Education*, 22(6), 802–820.

Zavala, M., & Simic-Muller, K. (Eds.). (2022). Special issue on antiracism in mathematics. *Teaching for Excellence and Equity in Mathematics*, 13(1).

Zavala, M., & Singwi-Ferrono, M. (2018). Balancing acts: Design-based mathematics for students living with trauma. In S. Crespo, S. Celedon-Pattichis, & M. Civil (Eds.), *Access & equity: Promoting high-quality mathematics (grades 3–5)* (pp. 49–66). National Council of Teachers of Mathematics.

Zavala, M., & Stoehr, K. (2019). From community exploration to social justice mathematics: How do mathematics educators help pre-service teachers make the move? In T. G. Bartell, C. Drake, A. Roth McDuffie, J. M. Aguirre, E. Turner, & M. Q. Foote (Eds.), *Transforming mathematics teacher education: An equity-based approach* (pp. 91–103). Springer Nature.

Zeichner, K. M., & Liston, D. P. (2014). *Reflective teaching: An introduction* (2nd ed.). Routledge, Taylor & Francis Group.

INDEX

**JENNIFER M. BAY-WILLIAMS,
JOHN J. SANGIOVANNI, ROSALBA SERRANO,
SHERRI MARTINIE, JENNIFER SUH**

Because fluency is so much more
than basic facts and algorithms

Grades K–8

**KAREN S. KARP,
BARBARA J. DOUGHERTY,
SARAH B. BUSH**

A schoolwide solution for students'
mathematics success

Elementary, Middle School, High School

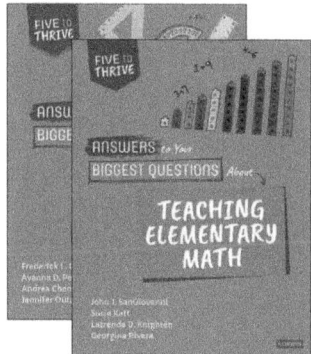

**JOHN J. SANGIOVANNI, SUSIE KATT,
LATRENDA D. KNIGHTEN, GEORGINA RIVERA,
FREDERICK L. DILLON, AYANNA D. PERRY,
ANDREA CHENG, JENNIFER OUTZS**

Actionable answers to your most
pressing questions about teaching
elementary and secondary math

Elementary, Secondary

**SARA DELANO MOORE,
KIMBERLY RIMBEY**

A journey toward making
manipulatives meaningful

Grades K–3, 4–8

A SAGE Publishing Company

Helping educators make the greatest impact

CORWIN HAS ONE MISSION: to enhance education through intentional professional learning.

We build long-term relationships with our authors, educators, clients, and associations who partner with us to develop and continuously improve the best evidence-based practices that establish and support lifelong learning.